The Structure
of Locally Compact
Abelian Groups

MONOGRAPHS AND TEXTBOOKS IN
PURE AND APPLIED MATHEMATICS

63. *W. L. Voxman and R. H. Goetschel,* Advanced Calculus: An Introduction to Modern Analysis (1981)
64. *L. J. Corwin and R. H. Szczarba,* Multivariable Calculus (1982)
65. *V. I. Istrățescu,* Introduction to Linear Operator Theory (1981)
66. *R. D. Järvinen,* Finite and Infinite Dimensional Linear Spaces: A Comparative Study in Algebraic and Analytic Settings (1981)
67. *J. K. Beem and P. E. Ehrlich,* Global Lorentzian Geometry (1981)
68. *D. L. Armacost,* The Structure of Locally Compact Abelian Groups (1981)
69. *J. W. Brewer and M. K. Smith, editors,* Emmy Noether: A Tribute to Her Life and Work (1981)

Other Volumes in Preparation

The Structure of Locally Compact Abelian Groups

D. L. Armacost
AMHERST COLLEGE
AMHERST, MASSACHUSETTS

MARCEL DEKKER, INC.   New York and Basel

Library of Congress Cataloging in Publication Data

Armacost, D. L. (David L.), [date]
  The structure of locally compact abelian groups.

  (Monographs and textbooks in pure and applied
mathematics ; 68)
  Includes bibliographical references and index.
  1. Locally compact Abelian groups. I. Title.
II. Series.
QA387.A75   512'.2   81-12527
ISBN 0-8247-1507-1   AACR2

MARCEL DEKKER, INC.

270 Madison Avenue, New York, New York 10016

Current printing (last digit):
10 9 8 7 6 5 4 3 2 1

PRINTED IN THE UNITED STATES OF AMERICA

Preface

The study of locally compact abelian (LCA) groups, the foundations of which were laid during the thirties by Pontryagin, Weil, and others, is of considerable importance in that much of classical harmonic analysis can be carried out conveniently and elegantly on groups of this type. But if the subject were suddenly to be deprived of all its applicability to analysis, it would not on that account cease to invite further study. Indeed, from the point of view of algebra and topology alone, the theory of these groups displays a charm and elegance which some, myself certainly included, have found irresistible. And though by the frenzied standards of our times the subject is no longer young, it still possesses the power to puzzle and delight.

My interest in these groups was first aroused by reading the elegant treatment to be found in the first volume of Hewitt and Ross' magnificent monograph *Abstract Harmonic Analysis* (which we regularly refer to by the symbol [HR]). In the years since the publication of this work (1963), and to a great extent because of its influence, the subject of LCA groups has received the attention of several investigators. Some years ago it occurred to me that the time was ripe to gather together and present in a unified manner the various algebraic and topological results on LCA groups that had been steadily appearing in the literature, as well as some findings (several of L. C. Robertson in particular) that were unpublished but deserved to be more widely known. It was not my intention to begin at the beginning, since the foundations are laid out so clearly in [HR] and, more recently, in the little book by Sidney Morris (see the references). Instead, I have brought together in a preliminary chapter all the necessary background information, drawn largely from [HR] and from Fuchs' monumental treatise on abelian groups [see Fuchs (1970, 1973)]. The function of this chapter is largely to establish notation and to provide a convenient source of references; a serious student of the subject will still doubtless make frequent use of [HR] and Fuchs' book.

A word about organization is in order. Before writing anything, I spent several years pondering what to present and how to mold the varied elements of the subject into a connected (or at least locally connected) narrative. To keep the size manageable, I resolved in the end to restrict myself primarily to algebraic and topological questions; as a result, measure theory, despite its many fascinations, finds no place in my scheme. The arrangement finally adopted is certainly not the one that I would have considered at the outset to be the most natural. However, I wanted the presentation to build progressively, so that each chapter would make maximum use of its predecessors, while no chapter would depend in any significant way upon its successors. The somewhat episodic nature of the subject made this a difficult task, but after trying various possibilities, I found myself being led almost inevitably to the arrangement finally adopted. I have taken great pains to make proofs as clear as I could without becoming prolix. Although the arrangement of the material sometimes necessitates longer proofs than are found in the literature, it has just as frequently allowed me to streamline some lengthy arguments. Near the end of each chapter is a "Miscellanea" section. Here I have gathered results ranging from simple exercises to summaries of research papers, together with several open questions. I have tried to ascribe all results in the book to their originators. I have nevertheless left some results (especially in the Miscellanea sections) unreferenced; several of these are my own concoctions, while others are of uncertain provenance. It is pleasant to hope, but rash to believe, that even so small a book as this will be free of errors, and I will be grateful to readers who send me notice of any they may find.

I have written the book with a somewhat diverse audience in mind. It is my hope that mathematicians interested in abelian group theory, topological algebra, or abstract harmonic analysis will find the book interesting. I further venture the wish that, although I have consulted my own predilections in the choice of material, the book may prove useful as a reference and as a guide to the literature.

The size of the book is no measure of the debt I owe to those who have helped in its production. For their encouragement in this project I am grateful to the members of my family and to my colleagues (especially

Professors Duane Bailey and Norton Starr) in the Mathematics Department of Amherst College. I also am indebted to Amherst College in the institutional sense for granting me generous leave time for research and writing. It is a pleasure to thank Mrs. Helen N. Sullivan, who not only produced a superb typescript, but also, by her careful attention to detail, saved me from many a lapse. My wife Gretchen did signal service in matters of presentation, proofreading, and indexing, without which help I would have been sore pressed indeed. Many mathematicians have of course aided me directly or indirectly, but I have an especial debt of gratitude to express to the memory of the late Professor Karel de Leeuw of Stanford University, who guided my earliest efforts in research and whose untimely death deprived the mathematical world of one of its most vibrant spirits.

Amherst, Massachusetts   D. L. Armacost

Contents

IN MEMORIAM
KAREL DE LEEUW

Preliminaries

*"La dernière chose qu'on trouve en faisant
un ouvrage est de savoir celle qu'il faut
mettre la première."* [Pascal, *Pensées* (I)]

P.1 *Set Theory* If X is a set and $S \subseteq X$ we shall write S^c for the complement of S in X. The cardinal number of a set X will be denoted by $|X|$. When we say "almost all $x \in X$" we mean "all but finitely many $x \in X$." The set of integers will be denoted by Z, the positive integers by Z^+, and the nonnegative integers by Z^{+0}. The set of all primes will be written $\mathscr{P}$. We regularly use the axiom of choice without mention, but the continuum hypothesis will not be invoked without explicit comment. We use $\mathfrak{c}$ to denote the power of the continuum.

P.2 *Abelian Groups* The class of abelian groups (without any topology assumed) will be denoted by $\mathscr{A}$. We generally employ additive notation, but we switch to the multiplicative in some instances where convenient. A subgroup H of $G \in \mathscr{A}$ is said to be *proper* iff $\{0\} \subsetneqq H \subsetneqq G$. The order of an element $x \in G$ is written $o(x)$. If f is a homomorphism from $G \in \mathscr{A}$ to $H \in \mathscr{A}$ we write $\ker f$ for the kernel of f and $\operatorname{im} f$ for the image of f in H. We use $G \simeq H$ to indicate that G and H are isomorphic ($\cong$ stands for topological isomorphism, for which see P.15).

P.3 *Further Notation for Abelian Groups* Let G denote a group in $\mathscr{A}$.
(a) For $n \in Z^+$ we define $G_{(n)} = \{x \in G : nx = 0\}$ and $nG = \{nx : x \in G\}$. Clearly $G_{(n)}$ and nG are subgroups of G.

1

(b) We denote by $t(G)$ the torsion subgroup of G, that is, $t(G) = \cup_{n=1}^{\infty} G_{(n)}$. We say that G is a *torsion* group iff $G = t(G)$, while G is *torsion-free* iff $t(G) = \{0\}$. Clearly the quotient group $G/t(G)$ is torsion-free.

(c) If $p \in \mathscr{P}$ we say that G is a *p-group* iff each $x \in G$ has order a power of p.

(d) If S is a subset of G we write gp(S) for the smallest subgroup of G containing S. If $S = \{x\}$ we write gp(x) for gp($\{x\}$). If $\{H_i\}_{i \in I}$ is an indexed collection of subgroups of G we write $\sum_{i \in I} H_i$ for gp($\cup_{i \in I} H_i$).

(e) Let $G_1, \ldots, G_n$ be in $\mathscr{A}$. The cartesian product group (direct product) is written $G_1 \times \cdots \times G_n$. If $\{G_i\}_{i \in I}$ is an indexed collection of groups $G_i \in \mathscr{A}$, we write $\Pi_{i \in I} G_i$ for the direct product. The subgroup of all sequences (x_i) in $\Pi_{i \in I} G_i$ such that $x_i = 0$ for almost all $i \in I$ is written $\Pi_{i \in I}^{*} G_i$ and referred to as the *weak direct product* of the groups G_i. If each $G_i = G$ for some fixed G and $\mathfrak{m}$ is a cardinal number we write $G^{\mathfrak{m}}$ for $\Pi_{i \in I} G_i$ and $G^{\mathfrak{m}*}$ for $\Pi_{i \in I}^{*} G_i$.

P.4 Important Groups The most important groups $G \in \mathscr{A}$ for our purposes are as follows:

(a) The cyclic groups, namely, Z under addition and the n-element cyclic groups $Z(n) = Z/nZ$.

(b) The additive group Q of rational numbers.

(c) Let $p \in \mathscr{P}$ be fixed. By $Z(p^{\infty})$ we mean the group of rationals expressible in the form m/p^n under addition mod 1 [$Z(p^{\infty})$ may also be thought of as the multiplicative group of p^nth roots of unity.] It is a fact (Fuchs 1970, p. 16) that each proper subgroup of $Z(p^{\infty})$ has the form $Z(p^n)$, for which reason the groups $Z(p^{\infty})$ are often called *quasicyclic*.

(d) Free abelian groups are the groups $Z^{\mathfrak{m}*}$, where $\mathfrak{m}$ is a cardinal number. Every $G \in \mathscr{A}$ is a quotient of some free abelian group. Moreover, a subgroup of a free abelian group is again free abelian. [See Fuchs (1970, p. 74).] We also have Pontryagin's theorem: If every subgroup of finite rank (see P.5 below) of a countable torsion-free group $G \in \mathscr{A}$ is free abelian, then G is itself free abelian. [See Fuchs (1970, §19.1).]

(e) The Specker group $S = Z^{\aleph_0}$. It turns out that although each countable subgroup of S is free abelian, S is itself not free abelian. [See Fuchs (1970, Theorem 19.2).]

P.5 Independence and Rank A collection $\{x_1, \ldots, x_n\}$ of nonzero elements of $G \in \mathscr{A}$ is said to be *independent* iff whenever $n_1, \ldots, n_k$ are integers such that $n_1 x_1 + \cdots + n_k x_k = 0$ then $n_1 x_1 = \cdots = n_k x_k = 0$. An infinite collection X of elements in G is said to be independent iff each finite subcollection of X is independent. In any $G \in \mathscr{A}$ there is (by Zorn's lemma) a maximal independent subset containing only elements of infinite or prime power order, and it turns out that any two such subsets have the same cardinality, which we call the *rank* of G, written $r(G)$. Similarly, the *torsion-free rank* of G, written $r_0(G)$, is the cardinal number (uniquely determined) of any maximal independent subset of G containing only elements of infinite order. Finally, for each $p \in \mathscr{P}$, the *p-rank* of G, written $r_p(G)$, is the cardinal number (again, uniquely determined) of any maximal independent subset containing only elements of order a power of p. We have $r(G) = r_0(G) + \sum_{p \in \mathscr{P}} r_p(G)$. We note here that if G is torsion-free and $r(G)$ is infinite, then $r(G) = |G|$. See Fuchs (1970, §16).

P.6 *Some Decomposition Theorems* Let G denote a group in $\mathscr{A}$.

(a) Let G be a torsion group and let G_p denote the subgroup of elements having order a power of $p \in \mathscr{P}$. Then each $x \in G$ can be written uniquely as a sum of elements in the various G_p's: We have $G \simeq \Pi^*_{p \in \mathscr{P}} G_p$. [See Fuchs (1970, §8.4).]

(b) Let G be *finitely generated* [in our notation, $G = \mathrm{gp}(F)$ for some finite subset F of G]. Then $G \simeq \Pi^n_{i=1} C_i$, where $n \in Z^+$ and each C_i is cyclic. We note for future reference the fact that $G \in \mathscr{A}$ is finitely generated iff the subgroups of G satisfy the maximum condition. [See Fuchs (1970, §15.5).]

(c) Let G have *bounded order* [i.e., $G = G_{(n)}$ for some $n \in Z^+$]. Then $G \simeq \Pi^*_{i \in I} C_i$, where I is some index set and each C_i is finite cyclic. [See Fuchs (1970, §17.2).]

P.7 *Direct Summands* A subgroup H of $G \in \mathscr{A}$ is said to be a *direct summand* of G iff there is a subgroup K of G such that $G = H + K$ and $H \cap K = \{0\}$. In this case we write $G = H \dotplus K$. (We shall use the $\oplus$ notation in a more special sense, for which see P.16.)

P.8 *Essential Subgroups and the Socle* A subgroup H of a group $G \in \mathscr{A}$ is said to be *essential* iff $H \cap K \neq \{0\}$ whenever $K \neq \{0\}$ is a subgroup of G. The *socle* of G is the subgroup consisting of all elements of square-free order. The following fact (Fuchs 1970, Ex. 10 on p. 87) is easily proved and quite useful: A subgroup H of G is essential iff H contains the socle of G and G/H is a torsion group.

P.9 *Divisible Groups* Let G stand for a group in $\mathscr{A}$.

(a) We say that G is *divisible* iff $G = nG$ for all $n \in Z^+$. We say that G is *reduced* iff $\{0\}$ is the only divisible subgroup of G. Clearly, direct products and homomorphic images of divisible groups are divisible. If G is divisible, so is $t(G)$. The groups Q and $Z(p^\infty)$ are divisible. [See Fuchs (1970, §20).]

(b) We shall frequently employ the following fact: G is divisible iff G/H is infinite for each proper subgroup H of G. [This is Ex. 2 on p. 99 of Fuchs (1970). One argument: If $nG \subsetneqq G$ for some $n \in Z^+$ apply P.6(c) to G/nG.]

(c) Let H be a subgroup of G and let $f : H \to D$ be a homomorphism, where $D \in \mathscr{A}$ is divisible. Then there exists a homomorphism $\bar{f} : G \to D$ which extends f (Fuchs 1970, §21.1).

(d) Let G have a divisible subgroup D. Then D is a direct summand of G. Even more can be said: namely, if H is a subgroup of G such that $H \cap D = \{0\}$ then there is a subgroup K of G such that $K \supseteq H$ and $G = D + K$. [See Fuchs (1970, §21.2).]

(e) The structure theorem for divisible groups: If D is divisible then $D \simeq Q^{r_0(D)*} \times \Pi^*_{p \in \mathscr{P}}[Z(p^\infty)^{r_p(D)*}]$. [See Fuchs (1970, §23).] An important special case is $Q/Z \simeq \Pi^*_{p \in \mathscr{P}} Z(p^\infty)$. Note also that the additive group of real numbers has the form Q^{c*}.

(f) Each G has a unique *maximal divisible subgroup* $d(G)$. A consequence is that any $G \in \mathscr{A}$ can be written in the form $D \times N$, where D is divisible and N is reduced. See Fuchs (1970, §21).

(g) Any G has a *minimal divisible extension* (or *divisible hull*) D, i.e., G may be embedded in a divisible group D having the property that if E is a divisible subgroup of D such that $G \subseteq E$ then either $G = E$ or $D = E$. If D' is another such minimal divisible extension then there is an isomorphism from D onto D' leaving the elements of G fixed. This unique (up to isomorphism) minimal divisible extension of G will be denoted by G^*. We have $r_0(G^*) = r_0(G)$ and $r_p(G^*) = r_p(G)$ for each $p \in \mathscr{P}$. Also, a divisible group D containing G is a minimal divisible extension of G iff G is essential in D (that is, by P.8, G contains the socle of D and D/G is torsion). Finally, the following statement (which follows easily from the above) is sometimes convenient. Let $G \in \mathscr{A}$ be divisible and let H be a subgroup of G. Then there is a subgroup D of G such that $D \supseteq H$ and D is isomorphic to the minimal divisible extension of H. [See Fuchs (1970, §24).]

(h) It will be useful for us to know that there exist reduced groups G (necessarily not torsion-free) such that $\cap_{n=1}^\infty nG \neq \{0\}$. One such example is given in §24.44(c) of [HR]. Another may be found on p. 150 of Fuchs (1970).

P.10 *Pure Subgroups*   A subgroup H of $G \in \mathscr{A}$ is said to be *pure* in G iff for each $n \in Z^+$ we have $H \cap nG = nH$. Divisible subgroups are clearly pure. If H is a subgroup of G such that G/H is torsion-free then H is pure; the converse holds if we also assume that G is torsion-free. Direct summands are pure [see (b) below for a partial converse]. Chapter V of Fuchs (1970) gives a full account of pure subgroups. We cite especially for later use:

(a) Let H be a subgroup of G. If H is finite (resp. infinite) there is a pure subgroup K of G such that $K \supseteq H$ and $|K| \leq \aleph_0$ (resp. $|K| = |H|$). If in addition G is torsion-free, there is a pure subgroup K of G such that $K \supseteq H$ and $r(K) = r(H)$. [See Fuchs (1970, pp. 115–116).]

(b) Let H be a pure subgroup of G. If either H is of bounded order or G/H is a weak direct product of cyclic groups, then H is a direct summand of G. [See Fuchs (1970, §§27.5 and 28.2).]

P.11 *Indecomposable Groups* A group $G \in \mathscr{A}$ which is not torsion-free contains a direct summand of the form $Z(p^n)$ or $Z(p^\infty)$ (Fuchs 1970, §27.3). Hence if G is *indecomposable* (that is, $\{0\}$ and G are the only direct summands of G) it follows that either $G \simeq Z(p^n)$, $G \simeq Z(p^\infty)$, or else G is torsion-free.

P.12 *Topological Groups* We shall use, generally without comment, the basic results of topological group theory (such as the fact that an open subgroup of a topological group is necessarily closed). See [HR] for a good exposition. For us, "neighborhoods" are always open. The closure of a subset A of a topological group G is written $\bar{A}$ (its complement being written A^c). Definitions from general topology are as in [HR].

P.13 *LCA Groups* Our subject is the structure theory of Hausdorff locally compact abelian groups (LCA groups). The class of such groups will be denoted by $\mathscr{L}$. If $G \in \mathscr{L}$ then by G_d we mean G with the discrete topology, and we write $\mathscr{L}_d$ for the class of discrete LCA groups. Our basic reference for LCA groups is [HR]. Other useful sources are Bourbaki (1967), Guichardet (1968), Husain (1966), Morris (1977), Pontryagin (1966), Reiter (1968), Rudin (1962), and Weil (1941, 1951). Finally we mention that since $\mathscr{L}$ and $\mathscr{A}$ are coextensive considered only as classes of groups, we use in $\mathscr{L}$ all the notation established above for $\mathscr{A}$.

P.14 *New Groups from Old* (a) If $G \in \mathscr{L}$ and H is a subgroup of G, then $H \in \mathscr{L}$ iff H is closed in G (one direction is standard topology; for the other see §5.11 of [HR]).

(b) If H is a closed subgroup of $G \in \mathscr{L}$ then G/H with the quotient topology is also in $\mathscr{L}$ (§5.22 of [HR]).

(c) If $G_1, \ldots, G_n$ are in $\mathscr{L}$ then $G_1 \times \cdots \times G_n$, when endowed with the product topology, also belongs to $\mathscr{L}$. If $\{G_i\}_{i \in I}$ is an indexed collection of groups in $\mathscr{L}$ then $\Pi_{i \in I} G_i$ with the product topology is in $\mathscr{L}$ iff $G_i \in \mathscr{L}$ for all $i \in I$ and G_i is compact for almost all $i \in I$ (§6.4 of [HR]). Unless mention is made to the contrary, direct products of groups in $\mathscr{L}$ are assumed to have the product topology.

(d) Let $\{G_i\}_{i \in I}$ be an indexed collection of groups $G_i \in \mathscr{L}$ and let H_i be a compact open subgroup of G_i for each $i \in I$. The *local direct product* of the groups G_i with respect to the subgroups H_i consists (as a group) of the subgroup of elements $(x_i) \in \Pi_{i \in I} G_i$ such that $x_i \in H_i$ for almost all $i \in I$. This group is topologized in such a way that $\Pi_{i \in I} H_i$ (with its compact product topology) is an open subgroup. We obtain in this way an LCA group which we denote by $\mathrm{LP}_{i \in I}(G_i : H_i)$. (See §6.16 of [HR].)

P.15 *Homomorphisms* Let G and H be in $\mathscr{L}$. The set of all *continuous* homomorphisms from G to H is denoted by $\mathrm{Hom}(G,H)$. An *open* homomorphism $f \in \mathrm{Hom}(G,H)$ is one such that if U is open in G then $f(U)$ is open in H. A *proper* homomorphism $f \in \mathrm{Hom}(G,H)$ is one such that if U is open in G then $f(U)$ is open in im f, with its topology inherited from H (note that im f must be closed in H since locally compact). The groups G and H are said to be *topologically isomorphic* iff there exists a *topological isomorphism* f from G onto H, i.e., an $f \in \mathrm{Hom}(G,H)$ such that $f(G) = H$, f is one-one, and f is open. In this case we write $G \cong H$. [We note in passing that two groups in $\mathscr{L}$ may be homeomorphic and algebraically isomorphic without being topologically isomorphic; see, for example, Robertson (1967, §2.14).] Finally, we note the following important, if elementary, fact: If $f \in \mathrm{Hom}(G,H)$ is open and surjective, then $H \cong G/\ker f$. (See §5.27 of [HR].)

P.16 *Topological Direct Summands* Let H be a closed subgroup of $G \in \mathscr{L}$. We say that H is a *topological direct summand* of G iff there is a closed subgroup K of G such that $G = H + K$ (see P.7) and such that the map $\phi : H \times K \to G$ defined by $\phi(h,k) = h + k$ for each $(h,k) \in H \times K$ is a topological isomorphism. In this case we write $G = H \oplus K$.

P.17 *Two Important Subgroups* Let G denote a group in $\mathscr{L}$.

(a) We denote the connected component of 0 in G by $c(G)$. Note that $c(G)$ is a closed subgroup of G (§7.1 of [HR]).

(b) An element $x \in G$ is said to be *compact* iff x lies in some compact subgroup of G [equivalently, $\overline{gp(x)}$ is compact]. We write $b(G)$ for the set of all compact (or "bounded") elements of G. Then $b(G)$ is a closed subgroup of G (§9.10 of [HR]). It is to be observed that $b(G)$ plays a role in $\mathscr{L}$ analogous to that of $t(G)$ in $\mathscr{A}$. For $G \in \mathscr{L}_d$ we have $t(G) = b(G)$, but in general $t(G) \subsetneqq \overline{t(G)} \subsetneqq b(G)$.

(c) The following fact is frequently useful: The set $b(G) + c(G)$ is always an open subgroup of G [§9.26(a) of [HR]].

P.18 *Important LCA Groups* (a) The groups $Z(n)$, Z, Q, $Z(p^\infty)$. These all have the discrete topology. Indeed, all countable LCA groups are discrete (cf. §5.28 of [HR]).

(b) The additive group R of real numbers with the usual topology. We have $b(R) = \{0\}$ and $c(R) = R$. It is easy to verify that all proper closed subgroups of R are discrete; in fact, if $H \neq \{0\}$ is a proper closed subgroup of R then there exists $x \neq 0$ in R such that $H = gp(x)$, i.e., $H \cong Z$. More generally, every closed subgroup of R^n with $n \in Z^+$ has the form $R^a \times Z^b$ where a and b are in Z^{+0} and $a + b \leq n$ (§9.11 of [HR]).

(c) The circle T consisting of all complex numbers z such that $|z| = 1$ under multiplication and having the usual euclidean topology. T may also be realized as the additive group R/Z. T is a compact and connected group. It is easy to show that if H is a proper closed subgroup of T, then $H \cong Z(n)$ for some $n \in Z^+$. It is also clear that T has no proper connected subgroups.

(d) The p-adic numbers F_p and the p-adic integers J_p. These are special cases of a more general construction (which we shall not use) to be found in §10 of [HR]. Let $p \in \mathscr{P}$ be fixed and consider the set F_p of all doubly infinite sequences $\bar{x} = (x_n)_{-\infty}^{+\infty}$ with $x_n \in \{0, 1, \ldots, p - 1\}$ for all n and $x_n = 0$ for almost all negative n. Each $\bar{x}$ is thought of as a formal power series $\sum x_n p^n$, and addition is defined accordingly (see §10.2 of [HR]). Under this addition F_p becomes an infinite abelian group (§10.3 of [HR]).

For each $k \in Z$ set $L_k = \{(x_n) \in F_p : x_n = 0 \text{ for } n < k\}$. Then the L_k's are subgroups of F_p, and when they are used to define a topology in F_p by declaring them to be a base at 0, then F_p becomes a totally disconnected LCA group in which the L_k's are compact. We have $b(F_p) = F_p$, but F_p is not compact. See §10.5 of [HR].

There is also a natural multiplication in F_p which turns F_p into a field having multiplicative identity $\bar{u} = (x_n)$ with $x_0 = 1$ but $x_n = 0$ for $n \neq 0$ (§10.10 of [HR]). The field F_p has characteristic zero, so that as an additive group F_p is divisible and torsion-free.

It turns out [§10.16(a) of [HR]] that if H is a proper closed subgroup of

F_p then $H = L_k$ for some k. Hence each proper closed subgroup of F_p is open (a very rare phenomenon—see 1.6). It is also easy to see that the L_k's are mutually topologically isomorphic, whence all proper closed subgroups of F_p are topologically isomorphic (again rare—see 1.10).

The p-adic integers J_p are defined to be the subgroup L_0 of F_p. Thus J_p is a compact, totally disconnected, and torsion-free group. (In fact, J_p becomes an integral domain under the multiplication mentioned for F_p.) It is clear from the preceding paragraph that every proper closed subgroup of J_p is open. Moreover, if H is a proper closed subgroup of F_p, then $H \cong J_p$.

P.19 *The Character Group* Let $G \in \mathcal{L}$. A homomorphism $\gamma : G \to T$ is called a *character* of G. We set $\hat{G} = \mathrm{Hom}(G, T)$, the collection of all *continuous* characters of G. The set $\hat{G}$ becomes a group (called the *character group* or *dual group* of G) under the pointwise operation. The trivial character is usually written 1. If F is a compact subset of G and $r > 0$ is a real number we set $P(F, r) = \{\gamma \in \hat{G} : |\gamma(x) - 1| < r$ for all $x \in F\}$. Then with the sets $P(F, r)$ as a basis of open sets at 1, $\hat{G}$ becomes an LCA group (§23.15 of [HR]). If G is compact (resp. discrete) then $\hat{G}$ is discrete (resp. compact) (see §23.17 of [HR]). We have the following facts:

(a) The continuous characters of G separate the points of G, that is, if $x \neq y$ belong to G there exists $\gamma \in \hat{G}$ such that $\gamma(x) \neq \gamma(y)$ (§22.17 of [HR]).

(b) If $G = G_1 \times \cdots \times G_n$ then $\hat{G} \cong \hat{G}_1 \times \cdots \times \hat{G}_n$ in a natural way. In fact, if $\gamma \in \hat{G}$ there exist $\gamma_i \in \hat{G}_i$, $i = 1, \ldots, n$, such that for each $\bar{x} = (x_1, \ldots, x_n) \in G$ we have $\gamma(\bar{x}) = \gamma_1(x_1) \cdots \gamma_n(x_n)$. Similarly, if $\{G_i\}_{i \in I}$ is an indexed collection of compact groups in $\mathcal{L}$ we have $\hat{G} \cong \Pi_{i \in I}^* \hat{G}_i$. See §§23.18 and 23.21 of [HR] for the details.

(c) Let G be the local direct product of groups G_i in $\mathcal{L}$ with respect to the compact open subgroups H_i [see P.14(d)]. Then $\hat{G}$ is topologically isomorphic with the local direct product of the $\hat{G}_i$ with respect to certain compact open subgroups K_i. In fact, in notation of P.22 below we have $K_i = A(\hat{G}_i, H_i)$. See §23.33 of [HR].

(d) For any $G \in \mathcal{L}$ and $x \in G$ define $\phi_x : \hat{G} \to T$ by $\phi_x(\gamma) = \gamma(x)$ for all $\gamma \in \hat{G}$. Then $\phi_x \in \mathrm{Hom}(\hat{G}, T) = (\hat{G})\hat{}$. The map $\Phi : G \to (\hat{G})\hat{}$ defined by $\Phi(x) = \phi_x$ for all $x \in G$ is a topological isomorphism from G onto $(\hat{G})\hat{}$. This is the celebrated *Pontryagin duality theorem*, which allows us to identify G with $(\hat{G})\hat{}$. (See §24.8 of [HR].)

(e) For $G \in \mathcal{L}$ let $w(G)$ denote the least cardinal number of an open basis of G. Then $w(G) = w(\hat{G})$ (§24.14 of [HR]). In particular, if G is compact we have $w(G) = |\hat{G}|$.

P.20 *Character Groups of Important LCA Groups* We refer to the list in P.18.

(a) If $\gamma \in (Z(n))\hat{}$ then γ is completely determined by $\gamma(1) \in Z(n) \subseteq T$. Thus $(Z(n))\hat{} \cong Z(n)$ [§23.27(c) of [HR]]. Hence if $F \in \mathscr{L}$ is finite we deduce from P.6(b) and P.19(b) that $\hat{F} \cong F$.

Each $\gamma \in \hat{Z}$ is determined by $\gamma(1) \in T$. In this way we get a correspondence between $\hat{Z}$ and T which turns out to be a topological isomorphism, i.e., $\hat{Z} \cong T$ [see §23.27(b) of [HR]].

To the character group of Q we assign no other name than $\hat{Q}$. This important compact group may be described as a *solenoid* (§25.4 of [HR]), and computations of the characters of Q may be found in §25.5 of [HR]. We note for the record that $\hat{Q}$ is torsion-free and connected [see P.22(f) and P.28(b) below].

We also have $(Z(p^{\infty}))\hat{} \cong J_p$ (§25.2 of [HR]). We shall not write out here a description of the characters of $Z(p^{\infty})$ [which comes down to the familiar algebraic problem of describing the endomorphisms of $Z(p^{\infty})$], since we have no explicit need for it in the sequel.

(b) Each $\gamma \in \hat{R}$ has the form $\gamma(x) = \exp(irx)$ for some fixed $r \in R$. This correspondence is a topological isomorphism, so that $\hat{R} \cong R$ [see §23.27(e) of [HR]].

(c) Each $\gamma \in \hat{T}$ has the form $\gamma(t) = t^n$ for some fixed $n \in Z$. We have $\hat{T} \cong Z$ [cf. $\hat{Z} \cong T$ and P.19(d)]. [See §23.27(a) of [HR].]

(d) It turns out (again, we neither reproduce here nor require later the actual computations) that $\hat{F}_p \cong F_p$. We also have $\hat{J}_p \cong Z(p^{\infty})$ [cf. $(Z(p^{\infty}))\hat{} \cong J_p$ and P.19(d)].

P.21 *Extensions* Let $G \in \mathscr{L}$ and let H be a closed subgroup of G.

(a) If $\gamma \in \hat{H}$ there exists $\overline{\gamma} \in \hat{G}$ such that $\overline{\gamma}$ extends γ. [In fact if $x \neq 0$ belongs to $G \cap H^c$, we can find such a $\overline{\gamma}$ satisfying $\overline{\gamma}(x) \neq 1$. See §24.12 of [HR].]

(b) If $f \in \text{Hom}(H,R)$ there exists $\overline{f} \in \text{Hom}(G,R)$ such that $\overline{f}$ extends f (§24.36 of [HR]).

P.22 *Annihilators* Let S be a nonempty subset of $G \in \mathscr{L}$. The *annihilator* of S in $\hat{G}$ is the set $A(\hat{G},S) = \{\gamma \in \hat{G} : \gamma(x) = 1 \text{ for all } x \in S\}$.

(a) If H is the smallest closed subgroup of G containing S then $A(\hat{G},H) = A(\hat{G},S)$ [[HR], §23.24(a)].

(b) $A(\hat{G},S)$ is a closed subgroup of $\hat{G}$ [[HR], §23.24(c)].

(c) If $S \neq \{0\}$ then $A(\hat{G},S) \neq \hat{G}$ [[HR], §23.24(b), or use P.19(a)].

(d) If H is a closed subgroup of G then $(G/H)\hat{} \cong A(\hat{G},H)$ and $\hat{H} \cong \hat{G}/A(\hat{G},H)$ ([HR], §§23.25 and 24.5).

(e) A closed subgroup H of G is open (resp. compact) iff $A(\hat{G},H)$ is compact (resp. open) [cf. part (d)].

(f) $A(\hat{G},G_{(n)}) = n\hat{G}$ and $A(\hat{G},nG) = (\hat{G})_{(n)}$ ([HR], §24.22).

(g) $A(\hat{G},c(G)) = b(\hat{G})$ and $A(\hat{G},b(G)) = c(\hat{G})$ ([HR], §24.17).

In like fashion we may define for any nonempty subset S' of $\hat{G}$ the set $A(G,S') = \{x \in G : \gamma(x) = 1 \text{ for all } \gamma \in S'\}$. We have:

(h) If H is a closed subgroup of G then $H = A(G,A(\hat{G},H))$ ([HR], §24.10).

(i) By virtue of the duality theorem P.19(d) the statements (a) through (h) above have analogues for $A(G,\cdot)$ in place of $A(\hat{G},\cdot)$. For example, if H' is a closed subgroup of $\hat{G}$, then $H' = A(\hat{G},A(G,H'))$. In particular, any closed subgroup H' of $\hat{G}$ is the annihilator of a unique closed subgroup of G. Similarly, we have $A(G,n\hat{G}) = G_{(n)}$ and $A(G,c(\hat{G})) = b(G)$, etc.

(j) For any $\gamma \in \hat{G}$ we have $A(\hat{G}, \ker \gamma) = \overline{\text{gp}(\gamma)}$. [For we have $\overline{\text{gp}(\gamma)} = A(\hat{G},A(G,\text{gp}(\gamma))) = A(\hat{G},A(G,\{\gamma\})) = A(\hat{G}, \ker \gamma).$]

Note: Whenever in the sequel we say "dualize" for a proof, we mean that the duality theorem and the appropriate properties of this section are to be used.

P.23 *Adjoint Homomorphisms* Suppose $G_1 \in \mathscr{L}$, $G_2 \in \mathscr{L}$, and $f \in \text{Hom}(G_1,G_2)$. We define $f^*: \hat{G}_2 \to \hat{G}_1$ by $f^*(\gamma_2) = \gamma_2 \circ f$ for all $\gamma_2 \in \hat{G}_2$. We have $f^* \in \text{Hom}(\hat{G}_2,\hat{G}_1)$. Moreover $\ker f^* = A(\hat{G}_2, \text{im} f)$ ([HR], §24.38). Then:

(a) Under the usual identification of G_1 and $(\hat{G}_1)\hat{}$ we have $f = (f^*)^*$ [[HR], §24.41(a)]. Hence we find $\ker f = A(G_1, \text{im} f^*)$.

(b) f^* is one-one iff f has dense image, and f^* has dense image iff f is one-one [[HR], §24.41(b)].

(c) f is a topological isomorphism from G_1 onto G_2 iff f^* is a topological isomorphism from $\hat{G}_2$ onto $\hat{G}_1$ [[HR], §24.41(c)].

(d) If f is proper (see P.15) then so is f^*; moreover we have $\text{im} f^* = A(\hat{G}_1, \ker f)$ ([HR], §24.40).

(e) If $G_3 \in \mathscr{L}$ and $g \in \text{Hom}(G_2,G_3)$ then $(g \circ f)^* = f^* \circ g^*$ (straightforward computation).

P.24 *Four Results for Compact Groups* Let $G \in \mathscr{L}$ be compact.

(a) If G is torsion-free then $G \cong (\hat{Q})^{\mathfrak{m}} \times \Pi_{p \in \mathscr{P}} (J_p)^{n_p}$, where $\mathfrak{m}$ and the

n_p's are cardinal numbers [[HR], §25.8; this follows from P.22(f), P.8(e), and duality].

(b) If G is a torsion group then $G \cong \Pi_{i \in I} C_i$, where C_i is a finite cyclic group for each i in the index set I and there is a finite upper bound on the set $\{|C_i|: i \in I\}$ ([HR], §25.9).

(c) Let U be a neighborhood of 0 in G. Then there is a closed subgroup H of G such that $H \subseteq U$ and $G/H \cong T^n \times F$, where $n \in Z^{+0}$ and F is finite ([HR], §24.7).

(d) Let G be infinite. Then $|G| = 2^{|\hat{G}|}$. [A proof of this important result of Kakutani (1943) may be found in §24.47 of [HR].]

P.25 Monothetic Groups A group $G \in \mathscr{L}$ is said to be *monothetic* iff there exists $x \in G$ such that $G = \overline{gp(x)}$. Clearly G is monothetic iff there exists $f \in \text{Hom}(Z, G)$ having dense image. An important result [due to Weil (1941, 1951)] is: If G is monothetic then either $G \cong Z$ or else G is compact. [See Rudin (1962, §2.3.2), or, for a more general result, §9.1 of [HR].]

P.26 Compactly Generated Groups A group $G \in \mathscr{L}$ is said to be *compactly generated* iff $G = gp(F)$ for some compact subset F of G. We have:

(a) $G \in \mathscr{L}$ is compactly generated iff $G \cong R^n \times Z^m \times K$, where m and n are in Z^{+0} and K is compact ([HR], §9.8).

(b) If S is a compact subset of a group $G \in \mathscr{L}$ there is a compactly generated open subgroup U of G such that $S \subseteq U$ ([HR], §5.14).

P.27 Connectedness Properties Let G be any group in $\mathscr{L}$. We have:

(a) $G/c(G)$ is totally disconnected ([HR], §7.3).

(b) If G is totally disconnected and U is a neighborhood of 0 in G there is a compact open subgroup K of G such that $K \subseteq U$ ([HR], §7.7). This is often phrased: G has arbitrarily small compact open subgroups.

(c) $c(G)$ is the intersection of all the open subgroups of G ([HR], §7.8).

(d) If G is totally disconnected and H is a closed subgroup of G, then G/H is also totally disconnected. [[HR], §7.11. Note that "zero-dimensional" and "totally disconnected" are equivalent concepts in $\mathscr{L}$ (cf. §§4.8 and 7.7 of [HR]).]

(e) *Structure theorem for connected groups.* G is connected iff $G \cong R^n \times K$, where $n \in Z^{+0}$ and K is compact and connected ([HR], §9.14). In particular, if G is connected then G is divisible [see P.28(b) below].

P.28 Characters and Connectedness Let $G \in \mathscr{L}$. We have:

(a) If G is connected then each $\gamma \neq 1$ in $\hat{G}$ is surjective [for $\gamma(G)$ is a connected subgroup of T, whence by P.18(c), $\gamma(G) = T$]. (For the converse see 8.2 below.)

(b) If G is compact the following are equivalent: (i) G is connected, (ii) $\hat{G}$ is torsion-free, (iii) G is divisible [[HR], §24.25; (i) ⇔ (ii) follows from (a) and P.27(c) and (ii) ⇔ (iii) follows from P.22(f)]. Note also that (i) ⇔ (ii) is a special case of P.22(g). Also from P.22(g) we obtain a companion statement: If G is compact, then G is totally disconnected iff $\hat{G}$ is a torsion group.

(c) If G is totally disconnected then ker γ is open in G for each $\gamma \in \hat{G}$. [Let I be a neighborhood of $1 \in T$ containing no proper subgroup of T (e.g., $I = \{z \in T : |z - 1| < \sqrt{3}\}$). Let U be a neighborhood of 0 in G such that $\gamma(U) \subseteq I$. By P.27(b) there is a compact open subgroup K of G with $K \subseteq U$. Then $K \subseteq$ ker γ, so ker γ is open.] The converse is also true, but we postpone a proof until 8.8.

(d) More generally, if $\gamma \in \hat{G}$ and $\gamma(c(G)) = \{1\}$ then ker γ is open [use (c) and P.27(a)].

P.29 Structure Theorem for LCA Groups Any $G \in \mathscr{L}$ is topologically isomorphic with $R^n \times G_0$, where $n \in Z^{+0}$ and $G_0 \in \mathscr{L}$ has a compact open subgroup ([HR], §24.30). In particular G has a closed subgroup $V \cong R^n$ and a closed subgroup G_0 containing a compact open subgroup such that $G = V \oplus G_0$. [It is a fact that n is uniquely determined by G ([HR], §24.30); it can also be shown that G_0 is uniquely determined up to topological isomorphism—see Armacost and Armacost (1978).]

P.30 Open Homomorphisms We have two frequently used results, one elementary and the other a Baire category result. Let G and H belong to $\mathscr{L}$ and let $f \in \text{Hom}(G, H)$.

(a) f is open iff for every neighborhood U of 0 in G there is a neighborhood V of 0 in H such that $V \subseteq f(U)$. [The proof is straightforward; see, for example, Pontryagin (1966, p. 113).]

(b) *Pontryagin's open mapping theorem.* If G is σ-compact and f is surjective then f is open ([HR], §5.29). [This result is frequently applied in the special case that G is connected, or more generally, compactly generated (cf. P.27(e)).]

P.31 Minimal Divisible Extensions Let $G \in \mathscr{L}$ and let G^* be a minimal divisible extension of G as in P.9(g). Define a topology on G^* by using an open basis at 0 in G as an open basis at 0 in G^*. Then G^* becomes an LCA group [cf. [HR], §25.32(a) and note that clearly G^* is locally compact]. We shall always use this topology for a minimal divisible extension G^* of G. Since G is open in G^* it follows from P.9(g) that any two minimal divisible extensions of G having their topologies defined as above must be topologically isomorphic, so we may speak of *the* minimal divisible extension G^* of G. The following facts are frequently useful:

(a) If $b(G) = G$ then $b(G^*) = G^*$. [This follows from the fact (P.9(g)) that G^*/G is a torsion group.]

(b) If G is totally disconnected so is G^* [see P.27(c)].

(c) If $D \in \mathscr{L}$ is divisible and contains G as an open subgroup, then $D \cong G^*$ iff G contains the socle of D and D/G is torsion [P.9(g)].

P.32 *The Bohr Compactification* For any $G \in \mathscr{L}$ we define the *Bohr compactification* $\beta(G)$ of G to be the compact group $((\hat{G})_d)\hat{\ }$. See [HR], §§26.12 and 26.13 for an equivalent definition.

P.33 *Metrizability* As is well known, a Hausdorff topological group G is metrizable iff there is a countable neighborhood basis at the identity of G (see §8.3 of [HR]). Since compact metric spaces are second countable, we see that a compact group $G \in \mathscr{L}$ is metrizable iff $w(G) \leqq \aleph_0$, i.e. [by P.19(e)], iff $|\hat{G}| \leqq \aleph_0$. [More generally, it is true that $G \in \mathscr{L}$ is metrizable iff $\hat{G}$ is σ-compact; for a proof of this fact (which we shall not use) see Morris (1977, Theorem 29).]

1

Familiar Groups Characterized

"New things are made familiar and familiar things made new." [Samuel Johnson, *Lives of the Poets* (Pope)]

In abelian group theory several of the most familiar and important groups can be characterized by simple properties of their subgroups. For example, Z is the only infinite abelian group all of whose proper subgroups have finite index. Similarly, the groups $Z(p^\infty)$ are characterized among the infinite abelian groups by the property that all their proper subgroups are finite. [See p. 22 of Kaplansky (1969) for both statements.] In this chapter we shall investigate analogous properties of the closed subgroups of LCA groups.

We begin with the group T. As pointed out in P.18(c), all of the proper closed subgroups of T are finite. This property of T characterizes T among the nondiscrete groups, but we wish to prove a bit more. First, however, as frequently in this chapter, we shall need a preliminary result about abelian groups.

1.1 Lemma Let $G \in \mathscr{A}$ be infinite. The proper subgroups of G are all finite iff $G \simeq Z(p^\infty)$ for some $p \in \mathscr{P}$.

Proof: The lemma appears as Exercise 23 on p. 22 of Kaplansky (1969). One argument is as follows: If G has the property in question, then all the nontrivial quotients of G must be infinite, whence by P.9(b), G is divisible. A glance at the structure theorem P.9(e) now reveals that $G \simeq Z(p^\infty)$ for some $p \in \mathscr{P}$. For the converse see P.4(c). ∎

We can now give our characterization of T, which is drawn from Robertson and Schreiber (1968), as are 1.3 through 1.7.

1.2 ***Theorem*** Let $G \in \mathscr{L}$ be infinite. Then every proper closed subgroup of G is finite iff $G \cong T$ or $G \cong Z(p^{\infty})$ for some $p \in \mathscr{P}$. Hence T is the only nondiscrete LCA group all of whose proper closed subgroups are finite.

Proof: If each proper closed subgroup of G is finite, the same must be true for any closed subgroup of G, whence G can contain no copy of R. The structure theorem P.29 then shows that G must contain a compact open subgroup K. If $K \subsetneq G$ then K is finite, so G is discrete, whence by Lemma 1.1, $G \cong Z(p^{\infty})$ for some $p \in \mathscr{P}$. If on the other hand $K = G$ then G is compact, and every proper quotient of the discrete group $\hat{G}$ is finite (P.22). It follows that $\hat{G}$ is finitely generated, and the structure theorem P.6(b) leads us to the conclusion that $\hat{G} \cong Z$, whence $G \cong T$. The converse has already been pointed out, and the last statement is now evident. ∎

1.3 ***Corollary*** Let $G \in \mathscr{L}$ be infinite. Then every proper closed subgroup of G has finite index iff $G \cong Z$ or $G \cong J_p$ for some $p \in \mathscr{P}$.

Proof: Dualize Theorem 1.2. ∎

If we replace the condition "Every proper closed subgroup is finite" by the weaker condition "Every proper closed subgroup is discrete," one further group, namely R, immediately comes to mind. The reader who can find no others (except discrete groups, of course) will be consoled by the next corollary. [As mentioned above, we draw this result from Robertson and Schreiber (1968), but it actually follows immediately from an older result of Isiwata (1955), for which see 1.19.]

1.4 ***Corollary*** Let $G \in \mathscr{L}$ be nondiscrete. Then every proper closed subgroup of G is discrete iff $G \cong T$ or $G \cong R$.

Proof: If G has the property in question, then P.29 reveals that either $G \cong R$ or else G has a compact open subgroup K. We cannot have $K \subsetneq G$, since then G would be discrete. Thus G must be compact, whence by 1.2 we have $G \cong T$. The converse is clear. ∎

1.5 ***Corollary*** Let $G \in \mathscr{L}$ be noncompact. Then each quotient of G by a proper closed subgroup is compact iff $G \cong Z$ or $G \cong R$.

Proof: This is just the dual of 1.4. ∎

One of the most striking properties of the groups J_p and F_p is that all their proper closed subgroups are open [P.18(d)]. It is remarkable that this property characterizes them among the nondiscrete groups. This fact, due to Robertson and Schreiber (1968), was also independently announced (without proof) in Soundararajan (1971).

1.6 *Theorem* Let $G \in \mathscr{L}$ be nondiscrete. Then each proper closed subgroup of G is open iff $G \cong J_p$ or $G \cong F_p$ for some prime p.

Proof: If G has the property in question, we see from P.29 that G must have a compact open subgroup K. Now every proper closed subgroup of K must be open in K. Since K is compact, it follows that every proper closed subgroup of K has finite index in K, whence by 1.3 either K is finite or $K \cong J_p$ for some $p \in \mathscr{P}$. Since G is nondiscrete, K cannot be finite, so $K \cong J_p$. Let f be a topological isomorphism from K onto the open subgroup J_p of F_p. Since F_p is divisible we can extend f to a homomorphism $\bar{f}: G \to F_p$ [P.9(c)]. Since $\bar{f}$ is continuous on the open subgroup K we have $\bar{f} \in \mathrm{Hom}(G, F_p)$. Moreover, it follows from P.30(a) that $\bar{f}$ is an open mapping. Finally, $\bar{f}$ is one-one; for if not, $\ker \bar{f}$ would be a proper closed subgroup of G, whence by hypothesis $\ker \bar{f}$ would be open in G, so that $\{0\} = (\ker \bar{f}) \cap K$ would be open in K, a contradiction. In sum, $\bar{f}$ is a topological isomorphism from G onto an open subgroup of F_p, so by P.18(d) we have $G \cong F_p$ or $G \cong J_p$. The converse has already been noted. ∎

1.7 *Corollary* Let G be a noncompact group in $\mathscr{L}$. Then each proper closed subgroup of G is compact iff $G \cong Z(p^{\infty})$ or $G \cong F_p$ for some $p \in \mathscr{P}$.

Proof: Dualize Theorem 1.6. ∎

We can, following Robertson and Schreiber (1968), give a rather amusing (if unnecessary) proof that $\hat{F}_p \cong F_p$. Indeed, set $G = \hat{F}_p$. Since every proper closed subgroup of F_p is open we have dually that every proper closed subgroup of G is compact. Clearly G is noncompact and not a $Z(p^{\infty})$-group, so by Corollary 1.7, $G \cong F_q$ for some prime q. The fact that G contains an isomorphic copy of J_p $[\cong (F_p/J_p)\hat{\ }]$ shows that $q = p$, so that $\hat{F}_p \cong F_p$.

The groups F_p may be regarded as natural companions of R. In fact, F_p can be realized as the completion of Q under the p-adic metric, while R is of course the completion of Q under the usual euclidean metric. The p-adic integer groups J_p bear a similar relation to the group Z. We can now formulate a property which sets these four types of groups apart from all other infinite LCA groups. Indeed, if G is one of the four groups Z, R, J_p, or F_p, and if H_1 and H_2 are any two proper closed subgroups of G, then $H_1 \cong H_2$ (see P.18). We aim to prove the converse of this statement. First, however, two algebraic lemmas.

1.8 Lemma Let $G \in \mathscr{A}$ be torsion-free. If $H \simeq Z$ for each proper subgroup H of G, then $G \simeq Z$.

Proof: We see from P.9(e) that G is not divisible. Hence $nG \subsetneqq G$ for some $n \in Z^+$. Since then $nG \simeq Z$ we can write $nG = \text{gp}(nx)$ for some $x \in G$. Since G is torsion-free we have $G = \text{gp}(x)$, so $G \simeq Z$. ∎

1.9 Lemma Let $G \in \mathscr{A}$ be infinite. If $G/H_1 \simeq G/H_2$ whenever H_1 and H_2 are proper subgroups of G, then $G \simeq Z(p^\infty)$ for some $p \in \mathscr{P}$.

Proof: Since G is infinite we can find a proper subgroup H of G. If G/H is finite then all proper quotients of G are finite, and 1.3 applied to G with the discrete topology shows that $G \simeq Z$, which is impossible. Hence G/H and therefore all proper quotients of G are infinite, so G is divisible [P.9(b)]. A glance at P.9(e) then gives $G \simeq Z(p^\infty)$ for some p. ∎

We can now get to our result, which, along with 1.11 through 1.13 following, is from a paper of the author (Armacost 1970).

1.10 Theorem The following are equivalent for an infinite group $G \in \mathscr{L}$:
  (a) $H_1 \cong H_2$ whenever H_1 and H_2 are proper closed subgroups of G.
  (b) G is topologically isomorphic with one of the four groups Z, R, J_p, or F_p.

Proof: Assume (a). We see from P.29 that either $G \cong R$ or else G has a compact open subgroup K. In the latter case, suppose first that G is discrete. If G had an element of finite order then 1.1 would give $G \cong Z(p^\infty)$, an impossibility. Hence G is torsion-free, whence 1.8 gives $G \cong Z$. We are left with the case that G is nondiscrete, whence $K \neq \{0\}$. Assume that G is noncompact. Then $\{0\} \subsetneqq K \subsetneqq G$, whence by hypothesis all proper closed subgroups of G are compact, so $G \cong F_p$ for some $p \in \mathscr{P}$ by 1.7. If on the other hand G is compact, the discrete group $\hat{G}$ satisfies the conditions of 1.9, so $\hat{G} \cong Z(p^\infty)$, whence $G \cong J_p$ for some p. This proves (a) $\Rightarrow$ (b). The converse has already been noted. ∎

1.11 Corollary The only infinite groups in $\mathscr{L}$ which are topologically isomorphic with each of their proper closed subgroups are Z and the groups J_p.

   Proof: This follows immediately from Theorem 1.10. ∎

   We next observe that if H is a proper closed subgroup of $G = T$ or $G = R$, then $G/H \cong T$, whence $H = \ker \gamma$ for some $\gamma \in \hat{G}$. Likewise, if H is a proper closed subgroup of $G = J_p$ or $G = F_p$, then H is one of the subgroups L_k [P.18(d)], and it is easily checked that G/H is a discrete subgroup of $Z(p^\infty)$, so again $H = \ker \gamma$ for some $\gamma \in \hat{G}$. We now show that, besides the discrete groups, the four types of groups mentioned are the only ones in $\mathscr{L}$ with this property.

1.12 Proposition Let $G \in \mathscr{L}$ be nondiscrete. The following are equivalent:
   (a) Each proper closed subgroup of G has the form $\ker \gamma$ for some $\gamma \in \hat{G}$.
   (b) G is topologically isomorphic with one of the groups T, R, J_p, or F_p.

   Proof: Assume (a), and set $C = c(G)$. If $C \neq \{0\}$ then C has a proper closed subgroup H. [This is a special case of 1.23; in this case, however, the result may be quickly seen by using P.27(e) and P.28(b).] Then by hypothesis $H = \ker \gamma$ for some $\gamma \in \hat{G}$. Since by P.28(a) we have $\gamma(C) = T$, we certainly have $\gamma(G) = T$ as well, whence $G = C + \ker \gamma = C$. The message of this paragraph is then that G is either totally disconnected or connected.

   Assume now that G is totally disconnected. Then (a) and P.28(c) together imply that each proper closed subgroup of G is open, whence by 1.6, $G \cong J_p$ or $G \cong F_p$. If on the other hand G is connected, we see from P.27(e) that either $G \cong R$ or G is compact. If G is compact, let B be any proper subgroup of the discrete torsion-free group $\hat{G}$. Now $\hat{B} \cong G/A\,(G,B)$, and by (a) we have $A\,(G,B) = \ker \gamma$ for some $\gamma \neq 1$ in $\hat{G}$. But since G is connected, γ is an open mapping from G onto T [P.28(a) and P.30(b)], so $G/\ker \gamma \cong T$ (cf. P.15). Thus $\hat{B} \cong T$, or $B \cong Z$. Since B was an arbitrary proper subgroup of $\hat{G}$ we find from 1.8 that $\hat{G} \cong Z$, or $G \cong T$. Hence (a) $\Rightarrow$ (b). The converse has already been noted. ∎

1.13 Corollary Let $G \in \mathscr{L}$ be noncompact. If every proper closed subgroup of G is monothetic, then G is topologically isomorphic with one of the following: Z, R, $Z(p^\infty)$, or F_p.

Proof: This is just the dual of Proposition 1.12, but we write out the details. Set $L = \hat{G}$ and let H be any proper closed subgroup of L. Since $\hat{L} \cong G$ we know that $A(\hat{L}, H)$, being a proper closed subgroup of $\hat{L}$, has the form $\overline{\mathrm{gp}(\lambda)}$ for some $\lambda \in \hat{L}$. But by P.22(j) we have $A(\hat{L}, \ker \lambda) = \overline{\mathrm{gp}(\lambda)}$ too, so $H = \ker \lambda$. Since L is nondiscrete and H is an arbitrary proper closed subgroup of L, we find from Proposition 1.12 that L is topologically isomorphic with T, R, J_p, or F_p, whence the result follows by taking duals. ∎

For a related result see 5.31.

The circle T is such a special group that we might expect it to have many different properties each of which characterizes T. In 1.2 we showed that T is the only nondiscrete group in $\mathscr{L}$ all of whose proper closed subgroups are finite. Looking a bit closer, we observe that each proper closed subgroup of T has the form $T_{(n)}$ for some $n \in Z^+$. This property also characterizes T among the nondiscrete groups. We show a bit more than this, but we require two algebraic lemmas, which may be of some interest in elementary abelian group theory. They will also be useful in giving another characterization of the groups J_p.

1.14 *Lemma* Let $G \in \mathscr{A}$ be infinite. If each proper subgroup of G has the form $G_{(n)}$ for some $n \in Z^+$, then $G \simeq Z(p^\infty)$ for some $p \in \mathscr{P}$.

Proof: We first show that G cannot have bounded order. Indeed, if G had bounded order we see from P.6(a) and (c) that G would be a weak direct product of infinitely many cyclic groups of prime-power order, and only finitely many distinct primes could be involved. Hence for at least one of the primes p there would have to be at least two nontrivial cyclic direct summands A_p and B_p having p-power order. It is then plain that neither A_p nor B_p could be $G_{(n)}$ for any n. Hence G does not have bounded order. It follows from this and the hypothesis on G that G/H is infinite for each proper subgroup H of G. We conclude from P.9(b) and (e) that $G \cong Z(p^\infty)$ for some $p \in \mathscr{P}$. [We note that the $Z(p^\infty)$ groups do indeed have the property in question.] ∎

1.15 *Lemma* Let $G \in \mathscr{A}$ be infinite. If each proper subgroup of G has the form nG for some $n \in Z^+$, then $G \simeq Z$.

Proof: Let the property under consideration be referred to in this proof as property P. One verifies directly that any pure subgroup, and hence any direct summand, of G must also enjoy property P. Now a group of the form $Z(p^m) \times Z(p^n)$, with m and n in Z^+ and $p \in \mathscr{P}$, does not have property P. Therefore G can have no direct summand of this form. It follows as in the proof of 1.14 that G is not a group of bounded order. As a consequence, G must be torsion-free. For suppose $x \in G$ had order $m > 1$. Setting $H = \mathrm{gp}(x)$, we have $H = nG$ for some $n \in Z^+$, so $mng = 0$ for all $g \in G$, contradicting the fact that G does not have bounded order. Thus G is torsion-free, whence $G \simeq nG$ for each $n \in Z^+$. Therefore G with the discrete topology satisfies the hypothesis of 1.11, whence $G \simeq Z$. ∎

We can now give our characterization of T. The result is perhaps well known, though the author can cite no reference.

1.16 Proposition Let $G \in \mathscr{L}$ be infinite. If each proper closed subgroup of G has the form $G_{(n)}$ for some $n \in Z^+$ then $G \cong T$ or $G \cong Z(p^\infty)$ for some $p \in \mathscr{P}$. Hence T is the only nondiscrete group $G \in \mathscr{L}$ each of whose proper closed subgroups has the form $G_{(n)}$.

Proof: It is clear from P.28(a) that a connected LCA group cannot be a torsion group. Hence our hypothesis ensures that either $c(G) = G$ or $c(G) = \{0\}$. In the former case P.27(e) shows that G is compact. Hence by P.22(f) the (discrete) group $\hat{G}$ satisfies the condition of 1.15, so $\hat{G} \cong Z$ and hence $G \cong T$. If, on the other hand, G is totally disconnected, let K be a compact open subgroup of G. If K is finite then G is discrete, and 1.14 shows that $G \cong Z(p^\infty)$ for some $p \in \mathscr{P}$. Thus it only remains to show that K cannot be infinite. Indeed, it is clear that K enjoys the same property as G, that is, each proper closed subgroup of K has the form $K_{(n)}$. Hence if K were infinite its (discrete) dual would satisfy the condition of 1.15, whence $\hat{K} \cong Z$ and hence $K \cong T$. Since G is totally disconnected, this is a contradiction. ∎

At this point we observe that the groups J_p have the following property: If H is a proper closed subgroup of J_p then $H = nJ_p$ for some $n \in Z^+$ (indeed, n is a power of p). We can now prove that this property characterizes the groups J_p among the nondiscrete groups in $\mathscr{L}$. The result is stated in Soundararajan (1971) without proof.

1.17 Corollary Let $G \in \mathscr{L}$ be nondiscrete. If each proper closed subgroup of G has the form nG for some $n \in Z^+$ then $G \cong J_p$ for some $p \in \mathscr{P}$.

Proof: Let K be any proper closed subgroup of $\hat{G}$. Then [P.22(i)] $K = A(\hat{G},H)$ for some proper closed subgroup H of G. But $H = nG$ for some $n \in Z^+$, so $K = (\hat{G})_{(n)}$ [P.22(f)]. Thus $\hat{G}$ (being necessarily infinite) satisfies the condition of 1.16, so $\hat{G} \cong T$ or $\hat{G} \cong Z(p^\infty)$ for some $p \in \mathscr{P}$. Since G is not discrete, we have $G \cong J_p$. ∎

Now let p be a fixed prime. The groups $Z(p^n), Z(p^\infty), J_p$, and F_p are all closely interrelated. In fact, these four types constitute the smallest class in $\mathscr{L}$ containing the groups $Z(p^n)$ and closed under formation of duals and minimal divisible extensions. There is one notable closed subgroup property that all four types share—namely, the closed subgroups are totally ordered by inclusion. It is a remarkable fact, and one with which we close the main body of this chapter, that this property characterizes these groups. The proof given here is from Armacost (1970). [The author has since learned that the result was proved a bit earlier (Muhin 1967); it is also stated in Soundararajan (1971).]

1.18 Theorem The following are equivalent for a group $G \in \mathscr{L}$:

(a) The closed subgroups of G are totally ordered by inclusion.

(b) G is topologically isomorphic with one of the four types of groups $Z(p^n), Z(p^\infty), J_p$, or F_p.

Proof: Assume (a). First assume G discrete. [An argument for this case can be found on p. 302 of Fuchs (1960), but we include one here for completeness.] Since Z does not satisfy (a) it is clear that G can contain no copy of Z, i.e., G is a torsion group. If G is finite, P.6(b) implies that $G \cong Z(p^n)$ for some $p \in \mathscr{P}$ and $n \in Z$. If G is infinite, let H be a proper subgroup of G. Choosing any $x \in G$ with $x \notin H$ we must have $H \subseteq gp(x)$, and since $o(x) < \infty$ we see that H is finite. It now follows from 1.1 that $G \cong Z(p^\infty)$ for some $p \in \mathscr{P}$.

If now G is not discrete it follows from P.29 that G has a compact open subgroup K. Since K must satisfy (a), the same is true of the discrete group $\hat{K}$, whence by the first paragraph and duality we have either $K \cong Z(p^n)$ or $K \cong J_p$. The former case is impossible, since G is not discrete. But if the open subgroup K has the form J_p, condition (a) and the fact that the proper closed subgroups of J_p are all open together imply that the proper closed subgroups of G are all open too. We then invoke Theorem 1.6 to conclude that G has the form J_p or F_p. Thus (a) $\Rightarrow$ (b), and the converse being evident, we are finished. ∎

Miscellanea

1.19 (Isiwata 1955) Let $G \in \mathscr{L}$ be nondiscrete. If $\ker \gamma$ is discrete for each $\gamma \neq 1$ in $\hat{G}$ then $G \cong R$ or $G \cong T$.

1.20 Suppose that each proper closed subgroup of $G \in \mathscr{L}$ is compactly generated. Then either G is itself compactly generated or has the form $Z(p^\infty)$ or F_p for some $p \in \mathscr{P}$. [By P.29 either $G \cong R^n$ or else $c(G)$ is compact. In the latter case $b(G)$ is open (P.17(c)). If $b(G) = G$ use P.26(a) and 1.7. Otherwise, conclude that the discrete torsion-free group $G/b(G)$ is finitely generated. (For a converse see 7.7.)]

1.21 Let $G \in \mathscr{L}$ have the property that each of its proper closed subgroups is σ-compact. Then G is itself σ-compact. [If G is connected, use P.27(e); otherwise, find a proper closed subgroup U of G such that $|G/U| \leq \aleph_0$ (cf. 8.3).]

1.22 Let $G \in \mathscr{L}$ be nondiscrete and have the property that each of its proper compact subgroups is open. Then either (a) $G \cong R^n \times D$, where $n \in Z$ and D is discrete and torsion-free, or (b) G is an open subgroup of a group $Q^{m*} \times F_p$, where m is a cardinal and $p \in \mathscr{P}$. [Take the minimal divisible extension of G and use Mackey's theorem on the structure of divisible torsion-free groups (§25.33 of [HR]), of which we also give a proof later (see 9.6).]

1.23 (Ross 1965) Any infinite $G \in \mathscr{L}$ contains a proper closed subgroup. (Indeed, it is not hard to show that G must contain infinitely many such.)

1.24 Let $G \in \mathscr{L}$ be infinite.

   (a) If $\sum H_i$ is dense in G whenever $\{H_i\}_{i \in I}$ is an infinite collection of distinct closed subgroups of G, then either $G \cong T$ or else $G \cong Z(p^\infty)$ for some $p \in \mathscr{P}$. (This follows from 1.2 and 1.23.)

   (b) [Stated in Soundararajan (1971)] If each infinite family of closed subgroups has zero intersection then either $G \cong Z$ or $G \cong J_p$ for some $p \in \mathscr{P}$. [Dualize (a).]

1.25 (Armacost and Bruner 1973) Call a group $G \in \mathscr{L}$ an *S-group* iff no two distinct closed subgroups of G are topologically isomorphic.

   (a) Let $G = \mathrm{LP}_{p \in \mathscr{P}}(C_p : B_p)$ where, for each $p \in \mathscr{P}$, C_p is a discrete cyclic group of p-power order and B_p is a subgroup of C_p. Then G is an S-group.

(b) If $G \in \mathscr{L}$ is not connected, then G is an S-group iff G has the form given in (a). (See also 5.36.)

1.26 If every subgroup of a group $G \in \mathscr{L}$ is closed, then G is discrete. [If G is compact, show that G must be a torsion group and use P.24(b). The general case now follows from P.29.]

1.27 (Armacost 1976) Let $G \in \mathscr{L}$.

(a) The closed subgroups of G satisfy the minimum condition (that is, every descending sequence of closed subgroups of G is stationary after some point) iff $G \cong T^n \times D$, where $n \in Z^{+0}$ and D is discrete and finitely cogenerated. [The definition of "finitely cogenerated" may be found in §25 of Fuchs (1970); it is shown there that D is finitely cogenerated iff D is the product of a finite group and finitely many quasicyclic groups.]

(b) The closed subgroups of G satisfy the maximum condition (similar definition, of course) iff $G \cong A \times K$, where A is discrete and finitely generated and K is a product of finitely many groups J_p for various primes p.

(c) The closed subgroups of G satisfy both the minimum and maximum conditions iff G is finite.

1.28 (Khan 1980) A group $G \in \mathscr{L}$ is compactly generated iff the open subgroups of G satisfy the maximum condition. (Khan's paper also contains several other interesting results on chain conditions for various types of subgroups of LCA groups.)

1.29 If H and K are closed subgroups of $G \in \mathscr{L}$, then $H + K$ need not of course be closed (cf. §4.4 of [HR]). Muhin (1970) has described the class of all those groups $G \in \mathscr{L}$ such that $H + K$ is closed in G whenever H and K are closed subgroups of G. We refer the reader to Muhin's paper for the description, which is somewhat involved.

2

Topological *p*-Groups

"When will return the glory of your prime?" [Shelley, *A Lament*]

In the structure theory of LCA groups the subgroup of compact elements plays a part analogous to that played by the torsion subgroup in abelian group theory. But there are other ways to generalize the concept of torsion to the continuous case. In this brief chapter we generalize the notion of *p*-group (or *p*-primary group) from the class $\mathscr{A}$ to the class $\mathscr{L}$. Although the results of this chapter have some interest in their own right, our main object is to set the stage for the next chapter.

Most of our results are drawn from Vilenkin (1946), Braconnier (1948), and the much later paper Robertson (1967). (Vilenkin imposes a second-countability restriction on his LCA groups, but this restriction turns out, as usual, to be unnecessary in most cases.) The class of groups studied in this chapter can be defined in various ways, and we shall actually find it convenient to give two definitions, namely the "topological *p*-group" definition as in Vilenkin (1946) and Robertson (1967) (see 2.3 below) and the "*p*-primary group" definition as in Braconnier (1948) (see 2.9 below). The equivalence of these two definitions will be shown in Theorem 2.12. Several ideas about these groups go back to the work of Jacobson, Krull, and others [see the references in Braconnier (1948) and Vilenkin (1946).]

2.1 *Definition* Let $G \in \mathscr{L}$ and $p \in \mathscr{P}$. By the *p-component* G_p of G we mean the set of all $x \in G$ such that $\lim_{n \to \infty} p^n x = 0$. [$G_p$ must be distinguished from $G_{(p)} = \{x \in G : px = 0\}$.]

2.2 *Remarks and Examples* (a) G_p is clearly a subgroup of G. In general, however, G_p is not closed [see (d) below]; but if G is totally disconnected then G_p must be closed, the proof of which fact we postpone until Chapter 3 (see 3.8). Note also that $G_p \subseteq b(G)$.

(b) Clearly if $G \in \mathscr{L}_d$ then $G_p = \cup_{n=1}^{\infty} G_{(p^n)}$.

(c) It is straightforward to check that if G is J_p or F_p then $G_p = G$, while $G_q = \{0\}$ if $q \in \mathscr{P}$ is different from p.

(d) The p-component of T clearly contains all the p^nth roots of unity. Hence T_p is a dense subgroup of T and will therefore not be closed if in fact we have $T_p \subsetneqq T$. This turns out to be true and follows from an interesting general theorem in Robertson (1967) about connected groups. Actually, we prefer to show directly (see 2.6) that $t \in T_p$ iff t is already a p^nth root of unity, a fact which is not entirely obvious.

Following Vilenkin (1946) and Robertson (1967) we now define the title of this chapter.

2.3 *Definition* Let $p \in \mathscr{P}$ be fixed. A group $G \in \mathscr{L}$ is said to be a *topological p-group* iff $G_p = G$.

2.4 *Examples* (a) Clearly $G \in \mathscr{L}_d$ is a topological p-group iff G is a p-group.

(b) The groups J_p and F_p are torsion-free topological p-groups.

(c) The group $\Pi_{n=1}^{\infty} Z(p^n)$ is an example of a compact topological p-group having proper dense torsion subgroup.

The foregoing examples and a natural sense of propriety doubtless will lead the reader to expect that $G \in \mathscr{L}$ can be a topological p-group for at most one $p \in \mathscr{P}$ (unless, of course, $G = \{0\}$). This is not hard to prove, but we shall wait until 2.7(b).

2.5 *Proposition* Let $p \in \mathscr{P}$ be fixed. The class of topological p-groups is closed under the formation of closed subgroups, quotients by closed subgroups, finite direct products and, more generally, local direct products.

Proof: Straightforward. ∎

The reader may have conjectured at this point that the class is also closed under the operation of taking duals. This is true and appears as Corollary 2.13.

We now prove what we asserted in 2.2(d).

2.6 Lemma For any $p \in \mathcal{P}$, T_p consists precisely of the p^nth roots of unity.

Proof: Let us use additive notation, realizing T as the interval $[0, 1)$ with addition modulo Z. We have to show that if $\theta \in [0, 1)$ satisfies $\lim_{n \to \infty} p^n \theta = 0$ mod Z then $p^N \theta \equiv 0$ mod Z for some $N \in Z^+$. Set $\kappa = 1/3p$. Then there exists $N \in Z^+$ such that if $n \geq N$ we have $p^n \theta = m_n + \alpha_n$, where $m_n \in Z$ and $\alpha_n \in R$ satisfies $|\alpha_n| < \kappa$. Hence for $n \geq N$ we have $pm_n + p\alpha_n = p^{n+1}\theta = m_{n+1} + \alpha_{n+1}$, whence $|pm_n - m_{n+1}| = |p\alpha_n - \alpha_{n+1}| \leq (p + 1)\kappa = (p + 1)/3p < 2/3$. Since $pm_n - m_{n+1} \in Z$ we get $pm_n - m_{n+1} = 0$, whence we have $\alpha_{n+1} = p\alpha_n$ for all $n \geq N$. In particular we have $\alpha_{N+1} = p\alpha_N$, $\alpha_{N+2} = p\alpha_{N+1} = p^2\alpha_N$, . . . , and in general $\alpha_{N+r} = p^r\alpha_N$ for all $r \in Z^+$. Since $|\alpha_n| < \kappa$ for all $n \geq N$ we are forced to conclude that $\alpha_N = 0$, i.e., $p^N \theta = m_N \equiv 0$ mod Z, as desired. ∎

2.7 Remarks (a) It follows from Lemma 2.6 that if G is a topological p-group and $\gamma \in \hat{G}$ then $\gamma(x)$ has order a power of p for each $x \in G$. In particular no $\gamma \in \hat{G}$ can be surjective, whence [P.19(a) and P.28(a)] $c(G) = \{0\}$. By the duality theorem we see further that no continuous character on $\hat{G}$ can be surjective, so by the same reasoning $c(\hat{G})$ is trivial too. Hence if G is a topological p-group then both G and $\hat{G}$ are totally disconnected.

(b) It also follows from Lemma 2.6 that a nontrivial $G \in \mathcal{L}$ can be a topological p-group for at most one $p \in \mathcal{P}$.

We can now give a simple structure theorem [Vilenkin (1946) for topological p-groups, Braconnier (1948) for p-primary groups].

2.8 Proposition Let $G \in \mathcal{L}$ be compact and torsion-free. Then G is a topological p-group iff $G \cong J_p^{\mathfrak{m}}$ for some cardinal $\mathfrak{m}$.

Proof: This follows from P.24, 2.5, and 2.7(a). ∎

We note in passing that a companion theorem describing the compact torsion topological p-groups may be easily deduced from P.24(b).

We now come to Braconnier's definition.

2.9 Definition Let $p \in \mathcal{P}$ be fixed. A group $G \in \mathcal{L}$ is said to be *p-primary* iff for each $x \in G$ there exists $f \in \operatorname{Hom}(J_p, G)$ such that $f(\bar{u}) = x$, where $\bar{u} = (1, 0, 0, \ldots)$.

As mentioned earlier, we intend to show that the *p*-primary groups of Braconnier are the same as the topological *p*-groups of Vilenkin and Robertson. (We use both definitions, not only for historical reasons, but also because it is helpful to have a name for the property of 2.9). Useful in this and in other connections are the following two lemmas.

2.10 Lemma Let $p \in \mathscr{P}$. A group $G \in \mathscr{L}$ is *p*-primary iff for each $x \in G$ there exists $f \in \mathrm{Hom}(J_p, G)$ such that $x \in \mathrm{im}\, f$.

Proof: One direction is trivial. For the converse pick any $x \in G$. We must find $g \in \mathrm{Hom}(J_p, G)$ such that $g(\bar{u}) = x$. We use the fact (P.18) that J_p has a continuous multiplication making it into a ring with identity $\bar{u}$. By assumption there exists $f \in \mathrm{Hom}(J_p, G)$ and $\bar{x}_0 \in J_p$ such that $f(\bar{x}_0) = x$. Define $g : J_p \to G$ by $g(\bar{x}) = f(\bar{x}_0 \cdot \bar{x})$ for all $\bar{x} \in J_p$, where the " $\cdot$ " represents the multiplication referred to. Then $g \in \mathrm{Hom}(J_p, G)$ and $g(\bar{u}) = \bar{x}$. ∎

The next result answers the question: If $G \in \mathscr{L}$ and $\lim_{n \to \infty} p^n x = 0$ for some $x \in G$, what can we say about the monothetic group $\overline{\mathrm{gp}(x)}$? The answer is perhaps not unexpected.

2.11 Lemma Let $G \in \mathscr{L}$ and $p \in \mathscr{P}$. If $x \in G_p$ then either $\overline{\mathrm{gp}(x)} \cong Z(p^n)$ for some $n \in Z^{+0}$ or $\overline{\mathrm{gp}(x)} \cong J_p$.

Proof: We find it convenient notationally to use the duality theorem. We thus think of G as being $\hat{H}$ (where $H = \hat{G}$) and think of x as being $\gamma \in \hat{H}$. We are then to show that if $p^n \gamma \to 0$ (additive notation!) then $\overline{\mathrm{gp}(\gamma)}$ is topologically isomorphic with $Z(p^n)$ or J_p. Now for any $h \in H$ we observe that $\gamma(h) \in T_p$, so by 2.6, $\gamma(h)$ is a p^nth root of unity. Hence im γ is algebraically a subgroup of $Z(p^\infty)$, i.e., im $\gamma \simeq Z(p^n)$ for some $n \in Z^{+0}$ or im $\gamma \simeq Z(p^\infty)$. Further, since γ must annihilate $c(H)$ we see from P.28(d) that ker γ is open in H, so $H/\mathrm{ker}\, \gamma$ is discrete. Therefore we can write $H/\mathrm{ker}\, \gamma \cong Z(p^n)$ or $H/\mathrm{ker}\, \gamma \cong Z(p^\infty)$. Finally, by P.22(d) and (j) we have $(H/\mathrm{ker}\, \gamma)\hat{} \cong \overline{\mathrm{gp}(\gamma)}$, whence by taking duals we have $\overline{\mathrm{gp}(\gamma)} \cong Z(p^n)$ or $\overline{\mathrm{gp}(\gamma)} \cong J_p$, as desired. ∎

We can now give our main result, which is essentially from Robertson (1967), though arrived at by a different route.

2.12 Theorem The following are equivalent for $G \in \mathscr{L}$ and a fixed $p \in \mathscr{P}$:
(a) G is a topological *p*-group.
(b) G is *p*-primary.

(c) Each $\gamma \in \hat{G}$ has range contained in the subgroup of p^nth roots of unity in T.

(d) G has arbitrarily small compact open subgroups such that the corresponding quotients are discrete p-groups.

Proof: We prove the equivalences as follows: (a)$\Leftrightarrow$(b) and (a)$\Rightarrow$(c)$\Rightarrow$(d)$\Rightarrow$(a). Assume (a). By Lemma 2.11 we have $\overline{\mathrm{gp}(x)} \cong Z(p^n)$ or $\overline{\mathrm{gp}(x)} \cong J_p$ for each $x \in G$. Since $Z(p^n)$ is a quotient of J_p we conclude from Lemma 2.10 that G is p-primary, so (a)$\Rightarrow$(b). It is evident that (b)$\Rightarrow$(a). That (a)$\Rightarrow$(c) is the content of Remark 2.7(a). Now assume (c). Arguing as in 2.7(a) we see that G is totally disconnected, so G has arbitrarily small compact open subgroups [P.27(b)]. To get (d) we need only show that if K is *any* compact open subgroup of G, then G/K is a discrete p-group. But this is evident from the fact that if the discrete group G/K had an element of infinite order or of order q^n, where q is a prime different from p, we could use P.9(c) and the divisibility of Q or $Z(q^\infty)$ to construct a character $\bar{\gamma}$ of G/K with im $\bar{\gamma}$ not contained in the subgroup of p^nth roots of unity of T; this $\bar{\gamma}$ would then give rise to a $\gamma \in \hat{G}$ disallowed by (c). Thus (c)$\Rightarrow$(d). Finally, assuming (d), let $x \in G$ and let U be a neighborhood of 0 in G. Pick a (compact) open subgroup K of G with $K \subseteq U$ and G/K a discrete p-group. Then for sufficiently large n we have $p^n(x + K) = K$, i.e., $p^n x \in K \subseteq U$. Since U was an arbitrary neighborhood of $x \in G$ we have $\lim_{n \to \infty} p^n x = 0$, so (a) holds. ∎

We can now prove the assertion made after the proof of Proposition 2.5. The result is to be found in one form or another in Braconnier (1948), Robertson (1967), and Vilenkin (1946).

2.13 Corollary Fix $p \in \mathscr{P}$. A group $G \in \mathscr{L}$ is a topological p-group iff $\hat{G}$ is a topological p-group.

Proof: This follows immediately from the duality theorem and the implication (c)$\Rightarrow$(a) in the preceding result. ∎

In particular we note that a compact $G \in \mathscr{L}$ is a topological p-group iff $\hat{G}$ is a discrete p-group.

Miscellanea

2.14 [Based on Robertson (1968)] (a) Let $G \in \mathscr{L}$ and $p \in \mathscr{P}$. Let G_p^+ denote the set of all $x \in G$ such that $x \in \mathrm{im}\, f$ for some $f \in \mathrm{Hom}(F_p, G)$. Then

G_p^+ is a divisible subgroup of G and is contained in G_p. (To show that G_p^+ is a subgroup argue as in 2.10.)

   (b) A topological *p*-group G is divisible iff $G = G_p^+$.

2.15 (Robertson 1968) The minimal divisible extension G^* of a topological *p*-group G is again a topological *p*-group. [Show that $G \subseteq (G^*)_p^+$ (cf. 2.14) and use the minimality of G^*.]

2.16 Let G be a topological *p*-group. If q is a prime different from p then $qG = G$. Hence, if $p^n G = G$ for all $n \in Z^+$, then G is divisible. [First prove the result assuming G compact, using P.22(f) and 2.13. In the general case consider the compact group $\overline{\mathrm{gp}(x)}$ for each $x \in G$.]

2.17 Let H be a closed subgroup of $G \in \mathscr{L}$. If for some $p \in \mathscr{P}$ both H and G/H are topological *p*-groups then so is G. [First show that $c(G) = \{0\}$ and $b(G) = G$. Then pick any $x \in G$ and look at the compact totally disconnected monothetic group $\overline{\mathrm{gp}(x)}$. (A shorter proof could be given by invoking the Braconnier-Vilenkin theorem to be proved later (3.13). Cf. 3.17.)]

2.18 Let $G \in \mathscr{L}$. If $G_p = \{0\}$ for all $p \in \mathscr{P}$ then $G \cong R^n \times D$, where $n \in Z^{+0}$ and D is discrete and torsion-free. [First observe that if $G = \hat{Q}$ then $G_p \neq \{0\}$ (cf. 4.18 below). Then use P.24(a) for the compact case. Finish with P.29.]

2.19 (Armacost 1972) $G \in \mathscr{L}$ is a topological *p*-group iff the ranges of the continuous characters of G are totally ordered by inclusion.

2.20 (Braconnier 1948 and Vilenkin 1946) A compact second-countable topological *p*-group is a direct product of cyclic *p*-groups iff $t(G)$ is dense in G. [This is actually just a dualization, taking P.19(e) into account, of Prüfer's theorem that a countable *p*-group $A \in \mathscr{A}$ is a weak direct product of cyclic *p*-groups iff $\cap_{n=1}^{\infty} nA = \{0\}$. See Fuchs (1970, §17.3).]

2.21 (Braconnier 1948) Let G be a compact topological *p*-group. Then pG is an open subgroup of G iff G is a direct product of finitely many groups J_p and finitely many cyclic *p*-groups. [Braconnier (1948) states this in a different way, using a concept of rank; see Prop. 6 on p. 36 of his paper. We note that a very quick proof can be given by using known facts about finitely cogenerated groups, for which see Fuchs (1970, §25.1).]

2.22 Let G be a topological p-group with the following property: If K is any compact subgroup of G there exists $x \in G$ such that $K \subseteq \overline{\mathrm{gp}(x)}$. Then G is topologically isomorphic with one of the following: $Z(p^n)$, $Z(p^\infty)$, J_p, F_p. [This is a variant of a result of Krull (see Vilenkin (1946) for references). One argument is as follows. Use 2.11 to show that each compact subgroup of G has the form $Z(p^n)$ or J_p. Observe further that each closed compactly generated subgroup of G is compact (P.26(a)). Now use P.26(b) to show that every proper compact subgroup of G is open in G. The result may now be deduced from 1.22 in the nondiscrete case, while the discrete case is evident.]

2.23 (Braconnier 1948) Let G be a torsion-free topological p-group. If H_1 and H_2 are compact open subgroups of G then $H_1 \cong H_2$. [First show that if $D \in \mathscr{A}$ is a divisible p-group and F is a finite subgroup of D, then $D/F \cong D$. Hence show that the result holds if G has the form J_p^m. Then use 2.8 to get $H_1 \cong H_1 \cap H_2 \cong H_2$.]

2.24 The group $(Z(p))^m$, where $\mathfrak{m}$ is an infinite cardinal, is a topological p-group for which the conclusion of 2.23 still holds.

2.25 (Braconnier 1948) Let $G \in \mathscr{L}$ satisfy $G = G_{(p)}$. Then $G \cong (Z(p))^m \times (Z(p))^{n^*}$ for some cardinals $\mathfrak{m}$ and $\mathfrak{n}$. (A proof may also be found in [HR], §25.29.)

2.26 Let G be as in 2.24 and let G^* be the minimal divisible extension of G. Then G^* is a nondiscrete topological p-group which contains no proper dense subgroup. [Let H be a dense subgroup of G^* and pick any $x \in G^*$. Since $x + G$ is open in G^* we have $H \cap (x + G) \neq \varnothing$, so we can write $x = h - g$ for some $h \in H$ and $g \in G$. Thus $px = ph \in H$, and since x was arbitrary we have $pG^* \subseteq H$. But G^* is divisible, so $pG^* = G^*$, whence $H = G^*$.] [The story behind this example is as follows. In Dietrich (1972) the question was raised whether every nondiscrete $G \in \mathscr{L}$ contains a proper dense subgroup. A negative answer in the form of a counterexample, of which the above is a rewording, was supplied by Rajagopalan and Subrahmanian (1976), who also give necessary and sufficient conditions for $G \in \mathscr{L}$ to contain a proper dense subgroup. Later, proceeding in a different way, Khan (1980) showed that $G \in \mathscr{L}$ contains no proper dense subgroup iff (a) $c(G) = \{0\}$, (b) $b(G) = t(G)$, and (c) pG is open in G for all $p \in \mathscr{P}$. In particular (taking 2.16 into account) we see that a topological p-

group G has no proper dense subgroup iff G is a p-group such that pG is open in G.]

2.27 *Envoi* A great deal of further information on topological p-groups may be found in Braconnier (1948) and Vilenkin (1946). Among very many detailed structure theorems, the latter has several results describing necessary and sufficient conditions under which a topological p-group may be written as a local direct product of copies of one (or more) of the four "basic" topological p-groups $Z(p^n)$, $Z(p^\infty)$, J_p, F_p.

3

Topological Torsion Groups

> "*The old order changeth, yielding place to new.*" [Tennyson, *The Passing of Arthur*]

In the last chapter we generalized the class of p-groups in $\mathscr{A}$ to the class of topological p-groups in $\mathscr{L}$. What then should be the analogue in $\mathscr{L}$ of the class of groups in $\mathscr{A}$ whose elements have finite order? One way to proceed is to observe that if $G \in \mathscr{L}$ is a torsion group, then each $\gamma \in \hat{G}$ has range contained in the torsion subgroup of T, so that $c(G)$ and $c(\hat{G})$ are both trivial. We might, then, following Vilenkin (1946) and Braconnier (1948), consider the class of all $G \in \mathscr{L}$ such that both G and $\hat{G}$ are totally disconnected. Indeed, Vilenkin (1946) refers to such groups as "topologically periodic." We have seen in Remark 2.7(a) that a topological p-group is topologically periodic in this sense. Another way to proceed, which seems a bit more natural and which leads to the same class, is to observe that if $G \in \mathscr{L}$ is a torsion group then $\lim_{n \to \infty} n!x = 0$ for each $x \in G$. Following Robertson (1967), we make this a defining property.

3.1 Definition For $G \in \mathscr{L}$ set $G! = \{x \in G : \lim_{n \to \infty} n!x = 0\}$. Then G is said to be a *topological torsion group* iff $G! = G$.

The following is surely to be expected.

3.2 Proposition If $G \in \mathscr{L}$ is a topological p-group then G is a topological torsion group.

Proof: Picking any $x \in G$ and neighborhood U of 0 we must show that $n!x \in U$ for sufficiently large n. Since G is totally disconnected there is [P.27(b)] an open subgroup K of G with $K \subseteq U$. Now there exists $N \in Z^+$ such that $p^N x \in K$. But since p^N divides $n!$ for sufficiently large n and since K is a subgroup, we have $n!x \in K \subseteq U$ for all such n. ∎

3.3 Remarks (a) It is obvious that $t(G) \subseteq G!$ and that $G!$ is a subgroup of G for any $G \in \mathscr{L}$. But $G!$ need not be closed, as will appear from Lemma 3.4. Also see 3.23.

(b) For any $p \in \mathscr{P}$ we have $G_p \subseteq G!$. For if $x \in G_p$ we have by 2.11 either $\overline{\mathrm{gp}(x)} \cong Z(p^n)$ or $\overline{\mathrm{gp}(x)} \cong \overline{J_p}$. By 3.2 we have $n!x \to 0$ in $\overline{\mathrm{gp}(x)}$ and hence in G.

(c) It is straightforward to show that the statement of Proposition 2.5 holds for topological torsion groups in place of topological p-groups. In particular, the local direct product of topological p-groups for various primes p is a topological torsion group. The important Braconnier-Vilenkin theorem (3.13) asserts that all topological torsion groups can be represented in this way.

Just as in Chapter 2 we needed to examine T_p, we now turn our attention to $T!$. It is clear that $t(T) \subseteq T!$, and one might naturally expect, having seen Lemma 2.6, that the inclusion is actually an equality. This is not the case, however. Again, let us realize T as the interval $[0,1)$ with addition modulo Z. Set $\theta = e - 2 = \sum_{k=2}^{\infty} 1/k!$. Since e is irrational, we have $\theta \notin t(T)$, but we aim to show that $\theta \in T!$. For each $n \geq 2$ in Z let I_n be the integer $n!(1/2! + 1/3! + \cdots + 1/n!)$. Then for each such n we have $n!\theta = I_n + R_n$, where $R_n = n![1/(n+1)! + 1/(n+2)! + \cdots] = 1/(n+1) + 1/(n+1)(n+2) + \cdots < \sum_{j=1}^{\infty} (n+1)^{-j}$, a geometric series having sum $1/n$. Thus for $n \geq 2$ we have $n!\theta \equiv R_n$ mod Z with $0 < R_n < 1/n$, so $n!\theta \to 0$ mod Z.

The question now arises: Is this θ unusual, or could it be that T is actually a topological torsion group? We can answer this in the negative by modifying what we have so far. Observe that I_n is odd if n is even, so we may write $I_n = 2J_n + 1$, where $J_n \in Z^{+0}$ for all even $n \geq 2$. Now set $\phi = e/2 - 1$. For all even $n \geq 2$ we have $n!\phi = J_n + 1/2 + R_n/2$, which is just another way of saying that $n!\phi$ comes arbitrarily close to $1/2$ mod Z for large even n. Hence $\phi \notin T!$. We now sum up our findings.

3.4 Lemma $t(T) \subsetneq T! \subsetneq T$. ∎

We remark that the proper inclusion $T! \subsetneq T$ is also a consequence of Robertson's theorem referred to in 2.2(d). It would be of interest to find some explicit description of $T!$.

If G is a topological torsion group and $\gamma \in \hat{G}$, we have $\gamma(G) \subseteq T!$, whence by Lemma 3.4 the range of γ is not all of T. It follows as in 2.7(a) that both G and $\hat{G}$ are totally disconnected. We are now in a position to prove the converse, by which we see that the topological torsion groups and the topologically periodic groups are the same. The following is based on Robertson (1967).

3.5 Theorem The following are equivalent for any $G \in \mathscr{L}$:
  (a) G is a topological torsion group.
  (b) Neither G nor $\hat{G}$ has a continuous surjective character.
  (c) Both G and $\hat{G}$ are totally disconnected.
  (d) $\gamma(x)$ has finite order for each $x \in G$ and $\gamma \in \hat{G}$.
  (e) G has arbitrarily small compact open subgroups such that the corresponding quotients are discrete torsion groups.

Proof: We have already shown that (a) $\Rightarrow$ (b) $\Rightarrow$ (c). Now assume (c) and pick $x \in G$ and $\gamma \in \hat{G}$. By P.22(g) we have $b(G) = G$, whence the monothetic subgroup $K = \overline{\text{gp}(x)}$ is compact (P.25). Since K is totally disconnected, $\hat{K}$ is a torsion group [P.28(b)]. Thus if γ_0 denotes the restriction of γ to K, γ_0 has finite order, whence the order of $\gamma(x) = \gamma_0(x)$ is finite too, i.e., (c) $\Rightarrow$ (d). Now assume (d). Clearly (d) $\Rightarrow$ (b), and since (b) $\Rightarrow$ (c) has already been shown, we may assume that both G and $\hat{G}$ are totally disconnected, i.e., $c(G) = \{0\}$ and $b(G) = G$. To get (e), let U be any neighborhood of 0 in G. Since G is totally disconnected there is [P.27(b)] a compact open subgroup K of G such that $K \subseteq U$. Since $b(G) = G$ we have $b(G/K) = G/K$, that is, the discrete group G/K is a torsion group, so (d) $\Rightarrow$ (e). Finally, the proof that (e) $\Rightarrow$ (a) is straightforward (cf. the last part of the proof of Theorem 2.12). ∎

The following two results [both given in Robertson (1967)] are now evident.

3.6 Corollary Let $G \in \mathscr{L}$ be compact. Then G is a topological torsion group iff G is totally disconnected. ∎

3.7 Corollary $G \in \mathscr{L}$ is a topological torsion group iff $\hat{G}$ is a topological torsion group. ∎

Before pushing on to the Braconnier-Vilenkin theorem (3.13) let us pause to make a couple of observations. Let $G \in \mathscr{L}$ and let $p \in \mathscr{P}$. If $\lim_{n \to \infty} p^n x = 0$ for some $x \in G$, we see from 2.11 that $\overline{\mathrm{gp}(x)}$ is a topological p-group. It is very tempting to suppose by analogy that if $\lim_{n \to \infty} n! x = 0$ then $\overline{\mathrm{gp}(x)}$ is a topological torsion group. We have only to recall, however, the element $\theta \in T$ constructed for Lemma 3.4 to dispel this notion; for θ belongs to $T!$ but $\overline{\mathrm{gp}(\theta)} = T$, which is decidedly not a topological torsion group. Our other observation is that groups G satisfying $G! = \{0\}$ [called "topologically torsion-free" in Robertson (1967)] are rather rare. In fact, Robertson (1967) showed that $G \in \mathscr{L}$ is topologically torsion-free iff $G \cong R^n \times D$, where $n \in Z^{+0}$ and D is discrete and torsion-free. The reader will observe that this follows immediately from 2.18 and 3.3(b).

We have from P.6(a) that a discrete torsion group G is the weak direct sum of its p-components G_p. The Braconnier-Vilenkin theorem (3.13) states more generally that a topological torsion group G may be written as a local direct product of its subgroups G_p. Naturally, we wish to know that each G_p belongs to $\mathscr{L}$, or what is the same thing [cf. P.14(a)], that G_p is closed in G for each $p \in \mathscr{P}$. All we need for this is that $c(G) = \{0\}$. The result is from Robertson (1967).

3.8 Lemma Let $G \in \mathscr{L}$ be totally disconnected. Then G_p is a closed subgroup of G for each $p \in \mathscr{P}$.

Proof: We already know that G_p is a subgroup of G [2.2(a)]. Now let $x \in \overline{G_p}$ and let U be a neighborhood of 0 in G. By P.27(b) there is an open subgroup H of G with $H \subseteq U$. Since $x + H$ is a neighborhood of x there is $y \in G_p \cap (x + H)$. Now $p^n y \in H$ for sufficiently large n, whence $p^n x \in H \subseteq U$ for such n too. Since U was an arbitrary neighborhood of 0 we have $\lim_{n \to \infty} p^n x = 0$, i.e., $x \in G_p$, so G_p is closed. ∎

3.9 Remarks (a) Braconnier (1948) defines G_p as the set of all $x \in G$ for which there exists $f \in \mathrm{Hom}(J_p, G)$ such that $f(\bar{u}) = x$ (cf. 2.9). It follows from 2.10 and 2.11 that Braconnier's G_p is the same as ours (which is really Vilenkin's).

(b) As we have seen in the case $G = T$, G_p need not be closed. In fact, if G is compact and connected, it may be shown that G_p is dense in G for each $p \in \mathscr{P}$ (see 4.18). Hence the converse of Lemma 3.8 holds if we assume that G has compact identity component. The full converse obviously fails, as we see from $G = R$.

Now for any $G \in \mathscr{L}$ and $p \in \mathscr{P}$ set $G_p^{\#} = \overline{\sum_{q \neq p} G_q}$, the smallest closed subgroup of G containing the subgroups G_q for all primes $q \neq p$. We now investigate the decomposition of compact topological torsion groups. [Recall from 3.6 that a compact $G \in \mathscr{L}$ is a topological torsion group iff $c(G) = \{0\}$.]

3.10 Proposition Let $G \in \mathscr{L}$ be compact and totally disconnected. Then for each $p \in \mathscr{P}$ we have $G = G_p \oplus G_p^{\#}$ (see P.16 for notation). Thus if $x \in G$ and $p \in \mathscr{P}$ we can write x uniquely as $x = x_p + x_p^{\#}$, where $x \in G_p$ and $x_p^{\#} \in G_p^{\#}$. The map f defined by $f(x) = (x_p)_{p \in \mathscr{P}}$ is a topological isomorphism from G onto $\Pi_{p \in \mathscr{P}} G_p$.

Proof: Of course, the key to this is to dualize P.6(a) as applied to $\hat{G}$, but we are actually asserting a bit more. By 3.8 each G_p is closed, so it makes sense to speak of $G_p \oplus G_p^{\#}$. Setting $X = \hat{G}$ we have by P.6(a) that $X \cong \Pi_{p \in \mathscr{P}}^{*} X_p$, whence $\hat{X} \cong \Pi_{p \in \mathscr{P}} (X_p)\hat{\ }$. To simplify notation we set $H = \Pi_{p \in \mathscr{P}} (X_p)\hat{\ }$. By 2.13, $(X_p)\hat{\ }$ is a topological p-group for each $p \in \mathscr{P}$, whence we see that for any particular $p \in \mathscr{P}$ we have $H_p = (X_p)\hat{\ }$ and $H_p^{\#} = \Pi_{q \neq p} (X_q)\hat{\ }$ (with an obvious abuse of notation). Thus $H_p \cap H_p^{\#} = \{0\}$ and $H = H_p + H_p^{\#}$. Now by the duality theorem $G \cong H$. Letting g be a topological isomorphism from G onto H, it is straightforward to verify that $g(G_p) = H_p$ and $g(G_p^{\#}) = H_p^{\#}$, so $G_p \cap G_p^{\#} = \{0\}$ and $G = G_p + G_p^{\#}$, i.e., $G = G_p \dotplus G_p^{\#}$. Now the isomorphism ϕ of P.16 is evidently continuous, and since $G_p \times G_p^{\#}$ is compact, ϕ is open as well. Therefore we have $G = G_p \oplus G_p^{\#}$. Now consider the map f from G to $\Pi_{p \in \mathscr{P}} G_p$. It is evident that f is a homomorphism. Moreover, since $G = G_p \oplus G_p^{\#}$ the map $x \to x_p$ from G to G_p is continuous for each $p \in \mathscr{P}$, whence f is continuous. Now suppose that $x \neq 0$ in G. Then $g(x) \neq 0$ in $H = \Pi_{p \in \mathscr{P}} (X_p)\hat{\ }$, so some coordinate of $g(x)$ is nonzero, which means that for some $p \in \mathscr{P}$ we can write $g(x) = h_p + h_p^{\#}$, where $h_p \neq 0$ is in H_p and $h_p^{\#} \in H_p^{\#}$. It is then clear that $x_p = g^{-1}(h_p) \neq 0$. Thus if $x \neq 0$ some $x_p \neq 0$ too, which just says that f is one-one. Now since G is compact, im f is compact in $\Pi_{p \in \mathscr{P}} G_p$. But it is easily verified that im f contains the dense subgroup $\Pi_{p \in \mathscr{P}}^{*} G_p$ of $\Pi_{p \in \mathscr{P}} G_p$, so f is actually surjective. In sum, f is a continuous monomorphism from G onto $\Pi_{p \in \mathscr{P}} G_p$, and since G is compact, f must be open. Therefore f is a topological isomorphism. ∎

Proposition 3.10 tells us, among other things, that to each element x in a compact topological torsion group we can associate its "p-coordinate" x_p. But we can do this as well in an arbitrary topological torsion group.

3.11 Definition Let $G \in \mathscr{L}$ be a topological torsion group. For given $x \in G$ and $p \in \mathscr{P}$ take any compact subgroup K containing x. By 3.10 we can write $x = x_p + x_p^{\#}$ uniquely with $x_p \in K_p$ and $x_p^{\#} \in K_p^{\#}$. The element x_p does not depend on the choice of K, and we call x_p the *p-coordinate* of x.

Of course we must prove the assertion embodied in the definition. First observe that since $b(G) = G$ a compact subgroup K containing x can indeed be found. Now suppose that $x \in L$, where L is some compact subgroup of G satisfying $L \supseteq K$. Write $x = \bar{x}_p + \bar{x}_p^{\#}$ with $\bar{x}_p \in L_p$ and $\bar{x}_p^{\#}$ in $L_p^{\#}$. Then $x_p - \bar{x}_p = \bar{x}_p^{\#} - x_p^{\#} \in L_p \cap L_p^{\#} = \{0\}$ by 3.10, so $\bar{x}_p = x_p$. Finally, if K_1 and K_2 are both compact subgroups of G containing x, the preceding argument shows that the decompositions of x relative to K_1 and K_2 agree with the one relative to $K_1 \cap K_2$. Hence x_p is uniquely determined by x and the prime p.

The following is now immediate from Definition 3.11 and Proposition 3.10.

3.12 Lemma Let G be a topological torsion group. We have:
   (a) If $x \in G$ then $x = 0$ iff $x_p = 0$ for all $p \in \mathscr{P}$.
   (b) For any x and y in G and $p \in \mathscr{P}$ we have $(x + y)_p = x_p + y_p$. ∎

Now that we can break down in a natural way each element x of a topological torsion group into its p-coordinates x_p it is reasonable to suppose that we can somehow write G as a product of its p-components G_p. We cannot, of course, use the direct product in general, since the G_p's need not be compact [cf. P.14(c)]. The precise result is the following theorem discovered independently by Braconnier (1948) and Vilenkin (1946).

3.13 Theorem Let $G \in \mathscr{L}$ be a topological torsion group and let H be any compact open subgroup of G. Then H_p is a compact open subgroup of G_p for each $p \in \mathscr{P}$ and we have $G \cong \text{LP}_{p \in \mathscr{P}}(G_p : H_p)$.

Proof: Since G is totally disconnected it certainly contains some compact open subgroup H. Moreover, each G_p is closed (3.8), so $G_p \in \mathscr{L}$ for each $p \in \mathscr{P}$. Also $H_p = H \cap G_p$ is a compact open subgroup of G_p, so it makes sense to speak of $\text{LP}_{p \in \mathscr{P}}(G_p : H_p)$. Now define $f : G \to \text{LP}_{p \in \mathscr{P}}(G_p : H_p)$ by $f(x) = (x_p)$ for each $x \in G$, where x_p is as in Definition 3.11. We claim that f is the desired topological isomorphism.

First we must verify that f actually carries each x into the local direct product, i.e., that $x_p \in H_p$ for almost all $p \in \mathscr{P}$. To this end, observe that the discrete group G/H is a torsion group. Hence if $x \in G$ there exists $m \in Z^+$ such that $mx \in H$. By 3.12(b) we have $mx_p = (mx)_p \in H_p$. But the discrete group G_p/H_p is a p-group, so the coset $x_p + H_p$ has finite order $p^{n(p)}$, where $n(p) \in Z^{+0}$ for each $p \in \mathscr{P}$. Since $m(x_p + H_p) = H_p$ we find that $p^{n(p)}$ divides m for all $p \in \mathscr{P}$, whence $n(p) = 0$ for almost all $p \in \mathscr{P}$. But this just says that $x_p \in H_p$ for almost all $p \in \mathscr{P}$, as we wished to show.

That f is a monomorphism is the substance of Lemma 3.12. Moreover, Proposition 3.10 says that the restriction of f to the compact open subgroup H is a topological isomorphism onto the compact open subgroup $\Pi_{p \in \mathscr{P}} H_p$ of $\mathrm{LP}_{p \in \mathscr{P}}(G_p : H_p)$. Therefore f is both continuous and open [cf. P.30(a)]. The surjectivity of f follows from the fact that im f is an open (and hence closed) subgroup of $\mathrm{LP}_{p \in \mathscr{P}}(G_p : H_p)$ which contains the dense subgroup $\Pi^*_{p \in \mathscr{P}} G_p$. [Alternatively, one could argue directly that im f clearly contains each factor G_p of $\mathrm{LP}_{p \in \mathscr{P}}(G_p : H_p)$; since each element of $\mathrm{LP}_{p \in \mathscr{P}}(G_p : H_p)$ can be written as the sum of an element of $\Pi_{p \in \mathscr{P}} H_p$ and finitely many elements drawn from the different G_p's, we have im $f = \mathrm{LP}_{p \in \mathscr{P}}(G_p : H_p)$.] In sum, f is a topological isomorphism from G onto $\mathrm{LP}_{p \in \mathscr{P}}(G_p : H_p)$. ∎

3.14 Remark In Theorem 3.13 the choice of the compact open subgroup H is apparently quite arbitrary. However, if K is another compact open subgroup of G we can show that $K_p = H_p$ for almost all $p \in \mathscr{P}$ (see 3.21), so that $\mathrm{LP}_{p \in \mathscr{P}}(G_p : H_p)$ and $\mathrm{LP}_{p \in \mathscr{P}}(G_p : K_p)$ are really the same sets.

Theorem 3.13 will find application throughout the rest of our work. For the present we content ourselves with a description of the torsion groups in $\mathscr{L}$. Recall from P.24(b) that a compact torsion group $G \in \mathscr{L}$ must have bounded order and can be written as a direct product of cyclic groups. We can now give a description, admittedly not as sharp as in the compact case, of an arbitrary torsion group $G \in \mathscr{L}$. The result is from Braconnier (1948).

3.15 Proposition Let $G \in \mathscr{L}$ be a torsion group. Then there exist distinct primes $p_1, \ldots, p_n$ such that $G \cong G_1 \times \cdots \times G_n \times D$, where G_i is a p_i-group and D is a discrete torsion group satisfying $D_{p_i} = \{0\}$ for $i = 1, \ldots, n$.

Proof: G is a topological torsion group, so by Theorem 3.13 we have $G \cong \mathrm{LP}_{p \in \mathscr{P}}(G_p : H_p)$ for some compact open subgroup H of G. Since H is compact we know from P.24(b) that $H_p = \{0\}$ for almost all $p \in \mathscr{P}$. It is now easy to verify that the local direct product can be rearranged as asserted. [By P.6(a), D can of course be further decomposed into its p-components.] ∎

3.16 Corollary The following are equivalent for $G \in \mathscr{L}$:
  (a) Both G and $\hat{G}$ are torsion groups.
  (b) G is a torsion group of bounded order.
  (c) There exist primes $p_1, \ldots, p_n$ such that $G \cong G_1 \times \cdots \times G_n$, where G_i is a p_i-group of bounded order for $i = 1, \ldots, n$.

Proof: We first note the easily proved fact [stated as §2.5.4 of Rudin (1962)] that a group $L \in \mathscr{L}$ is of bounded order iff $\hat{L}$ is of bounded order too. Now assume (a) and let K be a compact open subgroup of G. Then $A(\hat{G},K)$ is a compact subgroup of the torsion group $\hat{G}$, so by P.24(b), $A(\hat{G},K)$ has bounded order. But $A(\hat{G},K) \cong (G/K\hat{)}$, so taking $L = G/K$ we find that G/K has bounded order. But K has bounded order [P.24(b) again!], so G has bounded order, i.e., (a) $\Rightarrow$ (b). Assuming (b) we apply 3.15 to G and note that D must be a direct product of only finitely many different p-groups, whence we obtain (c). Finally, if (c) holds, G has bounded order, whence $\hat{G}$ does too. It follows a fortiori that (c) $\Rightarrow$ (a). ∎

Miscellanea

3.17 Let H be a closed subgroup of $G \in \mathscr{L}$. If H and G/H are topological torsion groups, then so is G.

3.18 The minimal divisible extension of a topological torsion group G is again a topological torsion group.

3.19 (Braconnier 1948) There exist topological torsion groups G and H such that $G_p \cong H_p$ for each prime $p \in \mathscr{P}$, but G and H are not topologically isomorphic. [Take $G = \Pi_{p \in \mathscr{P}} Z(p)$ and $H = \hat{G}$.]

3.20 Let G be a torsion-free topological torsion group. If H is an open subgroup of G then $|H| = |G|$. [Since G is not discrete (unless trivial) H is infinite. Find (P.10(a)) a pure subgroup K of G such that $K \supseteq H$ and $|K| = |H|$. Then show that $K = G$.]

3.21 Let $G \in \mathscr{L}$ be compact and totally disconnected and let U be an open subgroup of G. Then $G_p = U_p$ for almost all $p \in \mathscr{P}$. Hence if G is an arbitrary topological torsion group and H and K are compact open subgroups of G we have $H_p = K_p$ for almost all $p \in \mathscr{P}$. [By 3.13 we have $G \cong \text{LP}_{p \in \mathscr{P}}(G_p : U_p)$. If $U_p \subsetneqq G_p$ for infinitely many primes p, then G would have a discrete infinite quotient group.]

3.22 Let G_p and G'_p be topological p-groups containing compact open subgroups H_p and H'_p respectively for each $p \in \mathscr{P}$. Set $G = \text{LP}_{p \in \mathscr{P}}(G_p : H_p)$ and $G' = \text{LP}_{p \in \mathscr{P}}(G'_p : H'_p)$.

     (a) Suppose $f \in \text{Hom}(G, G')$. Let f_p be the restriction of f to G_p. Then if $\bar{x} = (x_p) \in G$ we have $f_p(x_p) \in H'_p$ for almost all $p \in \mathscr{P}$ and $f(\bar{x}) = (f_p(x_p)) \in G'$.

     (b) (Braconnier 1948) G and G' are topologically isomorphic iff for each $p \in \mathscr{P}$ there exists a topological isomorphism f_p from G_p onto G'_p such that $f_p(H_p) = H'_p$ for almost all $p \in \mathscr{P}$.

3.23 If $G \in \mathscr{L}$ is totally disconnected then $G! = b(G)$. Hence $G!$ is an open and closed subgroup of G. [For the first statement, observe that both $b(G)$ and $(b(G))\hat{\ }$ are totally disconnected. For the second recall P.17(c).] Is it true that for arbitrary $G \in \mathscr{L}$ we have $\overline{(G!)} = b(G)$?

4

Sufficiency Classes

"Satis quod sufficit" [Old saying]

The Pontryagin duality theorem renders the study of characters vitally important in the investigation of the structure of LCA groups. The fact that the circle T is used as the codomain for homomorphisms from a group $G \in \mathscr{L}$ is no accident. In fact, it can be shown that if instead of $\hat{G} = \mathrm{Hom}(G,T)$ we were to use some other "character" group $\mathrm{Hom}(G,H)$, where $H \in \mathscr{L}$ is different from T, we would not obtain the analogue of the duality theorem for all LCA groups (see §25.36 of [HR]). Nevertheless, the study of continuous homomorphisms from groups $G \in \mathscr{L}$ to certain other fixed groups $H \in \mathscr{L}$ is also useful. For example, the theory of "real characters" [that is, members of $\mathrm{Hom}(G,R)$] is important in the study of the Laplace transform and of distributions on LCA groups (see the papers of Mackey and Riss cited in [HR]); real characters are also useful in discussing arcwise connectedness (see Chapter 8). Suffice it to say for the present that the study of continuous homomorphisms from LCA groups G into such groups as R, Q, $\hat{Q}$, $Z(p^{\infty})$ and F_p will yield information about, and help to organize the description of, the structure of LCA groups.

  In this chapter we shall be interested primarily in such questions as "Are there sufficiently many $f \in \mathrm{Hom}(G,H)$ to separate the points of G?" rather than in the structure of $\mathrm{Hom}(G,H)$ as a group. Unless mention is made to the contrary, results presented in this chapter (before the Miscellanea section) are from a paper of the author (Armacost 1971b).

4.1 *Definition* Let $H \in \mathscr{L}$ be fixed. The *sufficiency class* $\mathscr{S}(H)$ of H is the class of all groups $G \in \mathscr{L}$ such that for each $x \neq 0$ in G there exists $f \in \mathrm{Hom}(G,H)$ such that $f(x) \neq 0$.

Thus by P.19(a) we have $\mathscr{S}(T) = \mathscr{L}$. It is easy to show, in fact, by using §25.31 of [HR] (which we give later as 6.16), that if $\mathscr{S}(H) = \mathscr{L}$ for some $H \in \mathscr{L}$, then $H \cong T \times H_0$ for some $H_0 \in \mathscr{L}$ (cf. 6.38). Our first aim is to describe $\mathscr{S}(H)$ for several important groups H. To do this, a dual concept, to be described after the next definition, is most useful.

4.2 *Definition* Let G and H be in $\mathscr{L}$. A (not necessarily closed) subgroup K of G is called an *H-subgroup* of G iff there exists $f \in \mathrm{Hom}(H,G)$ such that $K = f(H)$.

Thus, for example, $K = \{1, -1\}$ is a Z-subgroup of T. It is also a J_2-subgroup. The set of p^nth roots of unity in T is a $Z(p^\infty)$-subgroup, as well as an F_p-subgroup, of T.

4.3 *Definition* Let $H \in \mathscr{L}$ be fixed. The *dual sufficiency class* $\mathscr{S}^*(H)$ of H is the class of all groups $G \in \mathscr{L}$ whose H-subgroups generate a dense subgroup of G.

For example, R^n belongs to $\mathscr{S}^*(R)$ for any $n \in Z^+$. Similarly, $T \in \mathscr{S}^*(J_p)$ for any $p \in \mathscr{P}$, since the groups $Z(p^n)$, which are quotients of J_p, generate a dense subgroup of T. Finally, it is obvious that $\mathscr{S}^*(Z)$ is all of $\mathscr{L}$.

To justify the use of the word "dual" in Definition 4.3 we need the following fact. To simplify notation we use (G,H) for $\mathrm{Hom}(G,H)$ in the computations.

4.4 *Lemma* Let G and H be in $\mathscr{L}$. Then $A(\hat{G}, \cap_{f \in (G,H)} \ker f)$ is the smallest closed subgroup of $\hat{G}$ containing all the $\hat{H}$-subgroups of $\hat{G}$.

Proof: By P.22(h) we have, for each $f \in (G,H)$, $\ker f = A(G, A(\hat{G}, \ker f))$. Hence we have $\cap_{f \in (G,H)} \ker f = \cap_{f \in (G,H)} A(G, A(\hat{G}, \ker f)) = A(G, \cup_{f \in (G,H)} A(\hat{G}, \ker f))$ which by P.22(a) is $A(G, \sum_{f \in (G,H)} A(\hat{G}, \ker f))$. Taking annihilators in $\hat{G}$ and using P.22(i) we obtain

$$A\left(\hat{G}, \bigcap_{f \in (G,H)} \ker f \right) = \overline{\sum_{f \in (G,H)} A(\hat{G}, \ker f)} \tag{i}$$

On the other hand we have from P.23(a) that $\ker f = A(G, f^*(\hat{H})) = A(G, \overline{f^*(\hat{H})})$. Taking annihilators in $\hat{G}$ we get $A(\hat{G}, \ker f) = \overline{f^*(\hat{H})}$. Therefore $\sum_{f \in (G,H)} A(\hat{G}, \ker f) = \sum_{f \in (G,H)} \overline{f^*(\hat{H})} = \overline{\sum_{f \in (G,H)} f^*(\hat{H})}$. The last expression can be rewritten as $\overline{\sum_{g \in (\hat{H}, \hat{G})} g(\hat{H})}$, since by P.23(a) each $g \in \mathrm{Hom}\ (\hat{H}, \hat{G})$ is the adjoint of a unique f (namely g^*) in $\mathrm{Hom}\ (G, H)$. Therefore we have

$$\overline{\sum_{f \in (G, H)} A(\hat{G}, \ker f)} = \overline{\sum_{g \in (\hat{H}, \hat{G})} g(\hat{H})} \qquad \text{(ii)}$$

The proof is completed by combining formulas (i) and (ii). ∎

We can now indicate the dual relationship between sufficiency classes and dual sufficiency classes. The result, which is basic to much of this chapter, is from Moskowitz (1967).

4.5 Theorem For G and H in $\mathscr{L}$ we have $G \in \mathscr{S}(H)$ iff $\hat{G} \in \mathscr{S}^*(\hat{H})$.

Proof: Since $G \in \mathscr{S}(H)$ iff $\cap_{f \in (G,H)} \ker f = \{0\}$, the assertion follows immediately from Lemma 4.4. ∎

We mention in passing that Theorem 4.5, together with the duality theorem and the obvious equality $\mathscr{S}^*(Z) = \mathscr{L}$, can be combined to give a wonderfully quick (and entirely circular) "proof" of the deep result [P.19(a)] that $\mathscr{S}(T) = \mathscr{L}$.

We now prove a simple result that will enable us to determine the sufficiency classes of several important groups.

4.6 Lemma Let $H \in \mathscr{L}$ be divisible. If $G \in \mathscr{L}$ and $U \in \mathscr{S}(H)$ for each compactly generated open subgroup U of G, then $G \in \mathscr{S}(H)$.

Proof: Picking any $x \neq 0$ in G we find by P.26(b) a compactly generated open subgroup U of G which contains x. Then there exists $f \in \mathrm{Hom}(U, H)$ with $f(x) \neq 0$. By P.9(c) we may extend f to a homomorphism $\bar{f}: G \to H$. Since U is open f is automatically continuous, so we have $\bar{f} \in \mathrm{Hom}(G, H)$ and $\bar{f}(x) \neq 0$. Since $x \neq 0$ was arbitrary we have $G \in \mathscr{S}(H)$. ∎

The following result is well known (see §24.34 of [HR]).

4.7 Proposition $\mathscr{S}(\mathrm{R})$ consists of all $G \in \mathscr{L}$ such that $b(G) = \{0\}$.

Proof: Since $b(R) = \{0\}$ it is clear that $b(G) = \{0\}$ for any $G \in \mathscr{S}(R)$. Conversely, assume $b(G) = \{0\}$ and let U be any compactly generated subgroup of G. Then from P.26(a) we see that U has the form $R^n \times Z^m$, and since both R and Z belong to $\mathscr{S}(R)$, we have $U \in \mathscr{S}(R)$. By Lemma 4.6 we have $G \in \mathscr{S}(R)$. ∎

An *R*-subgroup of a group $G \in \mathscr{L}$ is called a *one-parameter* subgroup. In 1948 Mackey stated that an LCA group G is connected iff the subgroup generated by its one-parameter subgroups is dense in G. A proof of this appears in [HR, §25.20] (see p. 425 of [HR] for the reference to Mackey's paper). We can now recognize this as the dual of Proposition 4.7, as follows.

4.8 Theorem A group $G \in \mathscr{L}$ is connected iff $G \in \mathscr{S}^*(R)$.

Proof: By P.22(g), G is connected iff $b(\hat{G}) = \{0\}$, which is equivalent, by 4.7, to $\hat{G} \in \mathscr{S}(R)$. But since $\hat{R} \cong R$ we have by 4.5 and the duality theorem that $\hat{G} \in \mathscr{S}(R)$ iff $G \in \mathscr{S}^*(R)$. ∎

It was pointed out by Halmos (1944) that there exists a compact abelian group which is algebraically isomorphic with R, namely $\hat{Q}$. [For $\hat{Q}$, like R, is a divisible torsion-free group of cardinality $\mathfrak{c}$ (P.28(b), P.22(f), and P.24(d)). See also Hartman and Ryll-Nardzewski (1957) and Hawley (1971) for a discussion of compact topologies on R.] Thinking of $\hat{Q}$, then, as a kind of compact version of R, we might expect in light of Theorem 4.8 that a *compact* connected group G could be characterized by the property that its $\hat{Q}$-subgroups generate a dense subgroup of G. This turns out to be the case, as will follow from our description of $\mathscr{S}(Q)$.

4.9 Proposition $G \in \mathscr{L}$ belongs to $\mathscr{S}(Q)$ iff G is discrete and torsion-free.

Proof: Assume G discrete and torsion-free and pick $x \neq 0$ in G. Define $f : \text{gp}(x) \to Q$ by $f(nx) = n$ for all $n \in Z$. By P.9(c) we may extend f to $\bar{f} \in \text{Hom}(G, Q)$, so $G \in \mathscr{S}(Q)$. Conversely, if $G \in \mathscr{S}(Q)$ it is clear that $c(G) = b(G) = \{0\}$, whence by P.29, G is discrete and torsion-free. ∎

4.10 Corollary A compact group $G \in \mathscr{L}$ is connected iff the $\hat{Q}$-subgroups of G generate a dense subgroup of G.

Proof: Since G is connected iff the discrete group $\hat{G}$ is torsion-free, we have by 4.5 and 4.9 that G is connected iff $G \in \mathscr{S}^*(\hat{Q})$. ∎

We now investigate a class of groups which is of considerable importance.

4.11 Definition A group $G \in \mathscr{L}$ is said to be *densely divisible* iff G contains a divisible subgroup D such that $\bar{D} = G$.

In the terminology of P.9(f) we see that $G \in \mathscr{L}$ is densely divisible iff $\overline{d(G)} = G$. It was first proved by Robertson (1967) that $G \in \mathscr{L}$ is densely divisible iff $\hat{G}$ is torsion-free. Robertson's proof of this important result is based on his study of homogeneous groups, but we shall prove it by the method of sufficiency classes. We need some terminology.

4.12 Definition Let G and H be in $\mathscr{L}$. The union of the H-subgroups of G is called the *H-constituent* of G. The subgroup of G generated by the H-constituent is called the *H-component* of G.

The definition of the H-constituent is from Robertson (1968). We adopt the additional concept of H-component as being slightly more convenient, although in many cases the two sets are the same, as the following examples [based on Robertson (1968)] indicate. (See 4.32 for cases in which the H-component may be strictly larger than the H-constituent.)

4.13 Examples (a) From 2.11 we infer that the J_p-constituent of any $G \in \mathscr{L}$ is its p-component G_p. Since G_p is a subgroup of G [2.2(a)] we see that the J_p-constituent is the same as the J_p-component.
  (b) The F_p-constituent of $G \in \mathscr{L}$ is the subgroup G_p^+ of 2.14. Hence the F_p-constituent is the same as the F_p-component.
  (c) From P.9(e) and the fact that each $Z(p^\infty)$ group is a quotient of Q we see that the Q-component of $G \in \mathscr{L}$ is the maximal divisible subgroup $d(G)$ of G. Also note that, if $x \in G$ belongs to im f for some $f \in \mathrm{Hom}(G,Q)$, then there exists $g \in \mathrm{Hom}(G,Q)$ such that $g(1) = x$. It follows immediately that the Q-constituent of G is already a subgroup of G and therefore equals the Q-component of G.
  (d) The R-constituent of $G \in \mathscr{L}$ is of course just the union U of one-parameter subgroups of G. By an argument like that near the end of (c) above we see that U is a subgroup of G, whence the R-constituent and the R-component are the same. [It turns out (8.19) that U is the arcwise connected component of 0 in G.]
  (e) We remark generally that in the language of 4.12, if G and H belong to $\mathscr{L}$, then $G \in \mathscr{S}^*(H)$ iff G has dense H-component.

We see from 4.13(c) and (e) that if $G \in \mathscr{L}$ then G is densely divisible iff $G \in \mathscr{S}^*(Q)$. Therefore by 4.5, G is densely divisible iff $\hat{G} \in \mathscr{S}(\hat{Q})$. We prove Robertson's theorem by describing $\mathscr{S}(\hat{Q})$.

4.14 Proposition $G \in \mathscr{L}$ belongs to $\mathscr{S}(\hat{Q})$ iff G is torsion-free.

Proof: Since $\hat{Q}$ is torsion-free the same must be true for any $G \in \mathscr{S}(\hat{Q})$. Conversely, suppose G is torsion-free and let U be any compactly generated open subgroup of G. Since $\hat{Q}$ is divisible [P.28(b)] we need only show, by 4.6, that $U \in \mathscr{S}(\hat{Q})$. Now by P.26(a) we have $U \cong R^n \times Z^m \times K$, where m and n are in Z^{+0} and K is compact and torsion-free. To get $U \in \mathscr{S}(\hat{Q})$ we need only show that $R, Z,$ and K belong to $\mathscr{S}(\hat{Q})$, or, by 4.5, that $R, T,$ and $\hat{K}$ belong to $\mathscr{S}^*(Q)$. It is obvious that R and T belong to $\mathscr{S}^*(Q)$. Finally, the discrete group $\hat{K}$ is a weak direct product of copies of Q and of various $Z(p^\infty)$ groups [see P.24(a)]. Since both Q and $Z(p^\infty)$ clearly belong to $\mathscr{S}^*(Q)$ we have $\hat{K} \in \mathscr{S}^*(Q)$, which completes the proof. ∎

We can now prove Robertson's theorem.

4.15 Theorem A group $G \in \mathscr{L}$ is densely divisible iff $\hat{G}$ is torsion-free.

Proof: This follows immediately from Proposition 4.14 and the paragraph preceding it. ∎

4.16 Remarks (a) It follows from P.22(f) and Theorem 4.15 that $G \in \mathscr{L}$ is densely divisible iff nG is a dense subgroup of G for each $n \in Z^+$.

(b) A compact densely divisible group G is automatically divisible and hence connected. Indeed, $\hat{G}$ is torsion-free, and we need only apply P.28(b).

(c) It was pointed out by Moskowitz (1967) that if G is compactly generated then G is divisible iff $\hat{G}$ is torsion-free. [For by P.26(a) we may write $G \cong R^n \times Z^m \times K$, where $m, n \in Z^{+0}$ and K is compact. If G is densely divisible, the same is true of each quotient of G, whence $m = 0$ and K is densely divisible. It then follows from (b) that G is divisible (even connected).]

(d) There are several examples of densely divisible groups $G \in \mathscr{L}$ which are not divisible. One example, due to Freudenthal, may be found in the useful survey of Hartman and Ryll-Nardzewski (1957) and in §24.44 of [HR]. Another example, due to Robertson (1967), is of a different type. Let $\mathfrak{m}$ be an infinite cardinal and let D be the minimal divisible extension of $J_p^{\mathfrak{m}}$. Then $\hat{D}$ is densely divisible but not divisible. We omit the proof, but recommend it to the interested reader: See Proposition 2.7 of Robertson (1967). For a third example, which is perhaps the easiest to describe, let p be a fixed prime and let G be the local direct product of $\aleph_0$ copies of F_p with respect to the compact open subgroups J_p. If $\bar{u}$ is the element $(1, 0, 0, \ldots)$ in J_p then the sequence $\bar{x}$ with $\bar{u}$ in each entry belongs to G, but $\bar{x}$ cannot be divided by p without going outside the local direct product. Hence G is not divisible. Nevertheless, it is easy to see that G is densely divisible, either directly or by using P.19(c) to show that $\hat{G}$ is torsion-free. Note that G is σ-compact, since the compact open subgroup $J_p^{\aleph_0}$ has countable index in G. Hence even relatively "small" groups may be densely divisible without being divisible. It can be shown, however, that if $G \in \mathcal{L}$ has a dense divisible subgroup of finite rank, then G must be divisible. See 5.39(e). For a general result on densely divisible groups see 6.14.

Lemma 4.6 gives us some reason to hope that we might be able to determine $\mathcal{S}(H)$ for a variety of divisible groups H. Having already done this for $H = R$, Q, and $\hat{Q}$, we turn our attention to $Z(p^\infty)$ and F_p.

4.17 Proposition Let $p \in \mathcal{P}$ be fixed and let $G \in \mathcal{L}$. The following are equivalent:

(a) $G \in \mathcal{S}(Z(p^\infty))$.

(b) G is totally disconnected and every compact open subgroup of G is a topological p-group.

Proof: First assume that G is compact. If (a) holds then by 4.5, $\hat{G} \in \mathcal{S}^*(J_p)$. Hence the discrete group $\hat{G}$ must be a p-group, so by 2.13, G is a topological p-group. Since topological p-groups are totally disconnected [2.7(a)] we have (a) $\Rightarrow$ (b). Conversely, if (b) holds, then G itself must be a topological p-group, so $\hat{G}$ is a discrete p-group. Since each group $Z(p^n)$ is a quotient of J_p, we have $\hat{G} \in \mathcal{S}^*(J_p)$, whence $G \in \mathcal{S}(Z(p^\infty))$, so (b) $\Rightarrow$ (a).

Next observe that $Z \in \mathcal{S}(Z(p^\infty))$. This can be proved easily by a direct argument, or we can observe that since the p^nth roots of unity form a dense subgroup of T we have $T \in \mathcal{S}^*(J_p)$, whence $Z \in \mathcal{S}(Z(p^\infty))$.

Now let G be arbitrary. If (a) holds the total disconnectedness of $Z(p^\infty)$ shows that $c(G) = \{0\}$. The first paragraph shows that each compact (not necessarily open) subgroup of G is a topological p-group, so we have (a) $\Rightarrow$ (b). Conversely, if (b) holds, we let U be any compactly generated open subgroup of G. By P.26(a) we have $U \cong Z^m \times K$, where $m \in Z^{+0}$ and K is compact. Since Z^m is discrete and U is open in G, we see that K is topologically isomorphic with a compact open subgroup of G, so K is a topological p-group, whence by the first paragraph $K \in \mathscr{S}(Z(p^\infty))$. By the second paragraph Z^m also belongs to $\mathscr{S}(Z(p^\infty))$, so $U \in \mathscr{S}(Z(p^\infty))$. Since U was arbitrary and $Z(p^\infty)$ is divisible we have $G \in \mathscr{S}(Z(p^\infty))$ by 4.6, i.e., (b) $\Rightarrow$ (a). ∎

We observe that if $G \in \mathscr{L}$ is discrete and torsion-free, condition (b) of Proposition 4.17 is trivially satisfied, so $G \in \mathscr{S}(Z(p^\infty))$ for any $p \in \mathscr{P}$. It follows by duality that a compact connected group belongs to $\mathscr{S}^*(J_p)$ for all primes p, so by 4.13(a), G_p is dense in G for each prime p. It is a curious fact that a strengthened converse of this holds, as appears in (b) of the next result.

4.18 Corollary Let $G \in \mathscr{L}$. Then we have:
   (a) If G is compact and connected, then G_p is dense in G for each $p \in \mathscr{P}$.
   (b) If G_p and G_q are both dense in G for distinct primes p and q, then G is compact and connected.

Proof: We have already proved (a). Assuming the hypothesis of (b) we see by 4.5 and 4.13(a) that $\hat{G}$ belongs to $\mathscr{S}(Z(p^\infty))$ and $\mathscr{S}(Z(q^\infty))$. Therefore by 4.17, $\hat{G}$ is totally disconnected, so $\hat{G}$ contains a compact open subgroup H. By 4.17 we see that H is at once a topological p-group and a topological q-group, so $H = \{0\}$ by 2.7(b), i.e., $\hat{G}$ is discrete. The same argument shows that any finite subgroup of $\hat{G}$ is trivial, so $\hat{G}$ is torsion-free. Hence G is compact and connected. ∎

4.19 Proposition Let $p \in \mathscr{P}$ be fixed and let $G \in \mathscr{L}$. The following are equivalent:
   (a) $G \in \mathscr{S}(F_p)$.
   (b) G is totally disconnected and every compact open subgroup of G is a product (perhaps trivial) of copies of J_p.

Proof: Assume (a). Then clearly G is totally disconnected and torsion-free and can contain no copy of J_q for q a prime different from p. Hence it follows from P.24(a) that each compact subgroup of G is a product of copies of J_p, so (a) $\Rightarrow$ (b). Conversely, assume (b) and let U be any compactly generated open subgroup of G. As in the proof 4.17 we have $U \cong Z^m \times K$, where $m \in Z^{+0}$ and K is compact. Since K is topologically isomorphic with a compact open subgroup of G, K is a product of copies of J_p. Since both Z and J_p belong to $\mathscr{S}(F_p)$, we have $U \in \mathscr{S}(F_p)$. Thus by 4.6 we have $G \in \mathscr{S}(F_p)$, i.e., (b) $\Rightarrow$ (a). ∎

From 2.14 and 4.13(b) we see that a topological p-group G is divisible iff G equals its F_p-component. We likewise see that if G has dense F_p-component then G is densely divisible. By using Proposition 4.19 we can prove the converse. The result is from Robertson (1968). (A related result may be found in 4.33.)

4.20 Corollary Let $p \in \mathscr{P}$ be fixed. A topological p-group G is densely divisible iff $G \in \mathscr{S}^*(F_p)$.

Proof: One direction has already been shown. Conversely, assume that G is densely divisible. By 4.15 and 2.13, $\hat{G}$ is a torsion-free topological p-group. Then by 2.8 each compact subgroup (open or not) of G is a product of copies of J_p. Since $\hat{G}$ is totally disconnected we have $\hat{G} \in \mathscr{S}(F_p)$ by 4.19. Hence by 4.5, $G \in \mathscr{S}^*(F_p)$. ∎

The preceding result can be used to prove an important structure theorem about nonreduced groups. To obtain this theorem we need two preliminary results, the first of which [from Robertson (1968)] is useful in a variety of situations. [Our proof is different from Robertson's; for yet a third proof (exploiting properties of the adjoint map) see Armacost (1974).]

4.21 Proposition Let $p \in \mathscr{P}$ be fixed and let P denote either of the groups $Z(p^\infty)$ or F_p. Let f be a continuous monomorphism from P into a totally disconnected group $G \in \mathscr{L}$. Then f is a topological isomorphism from P onto a closed subgroup of G.

Proof: We show that $f(P)$ is a closed (and hence locally compact) subgroup of G. Since P is σ-compact the conclusion will then follow from P.30(b).

First, a preliminary observation. If $K \in \mathscr{L}$ is compact and totally disconnected and $m \in Z^{+0}$, then $\mathrm{Hom}(P, Z^m \times K) = \{0\}$. For if $g \in \mathrm{Hom}(P, Z^m \times K)$ then, since $b(P) = P$, we have $g(P) \subseteq \{0\} \times K$. But then $\overline{g(P)}$ is compact and densely divisible, so by 4.16(b), $\overline{g(P)}$ is connected. Since $c(K) = \{0\}$ we conclude that $g = 0$, as desired.

We now show that if x is any element in the complement $(f(P))^c$ of $f(P)$, then there is an open subset V of G such that $x \in V \subseteq (f(P))^c$. Let H be a compactly generated open subgroup of G which contains x [P.26(b)]. Since $c(G) = \{0\}$ we conclude from P.26(a) that $H \cong Z^m \times K$, where $m \in Z^{+0}$ and K is compact. Now let $L = f^{-1}(H)$. Since L is a closed (indeed, open) subgroup of P, we have either (1) $L = P$ or (2) L is compact (this is the "easy" half of Corollary 1.7). If (1) held, then f would belong to Hom(P, H), so $f = 0$ by the preceding paragraph. Since this is absurd, we are left with (2), whence $f(L)$ is a compact subgroup of H. [$f(L)$ cannot be all of H, of course, since $x \notin f(G)$.] Setting $V = H \cap (f(L))^c$ we see that V is an open subset of G satisfying $x \in V \subseteq (f(P))^c$. This shows that $f(P)$ is closed in G, so we are done. ∎

We remark on leaving this result that the total disconnectedness of G is certainly necessary. As to $Z(p^\infty)$, we have only to observe that $Z(p^\infty)$ can be injected onto a dense subgroup of T. As to F_p, we see from 4.14 that there exists $f \neq 0$ in Hom(F_p, $\hat{Q}$). Clearly f must be one-one, but f is evidently not a topological isomorphism onto its image.

The second of our two preliminary results follows from the first and is also from Robertson (1968).

4.22 Proposition Let G be a nontrivial topological p-group. If G is densely divisible then G contains a copy of F_p or $Z(p^\infty)$.

Proof: By 4.20 we have $G \in \mathscr{S}^*(F_p)$. In particular, there exists $f \neq 0$ in Hom(F_p, G). If f is one-one then the total disconnectedness of G and 4.21 together imply that $f(F_p) \cong F_p$. Otherwise ker f is an open subgroup of F_p [see P.18(d) or 1.6], so $F_p/(\ker f) \cong Z(p^\infty)$. Hence there is a monomorphism (automatically continuous) $g : Z(p^\infty) \to G$, whence (4.21 again) $g(Z(p^\infty)) = Z(p^\infty)$. ∎

We now can prove the theorem of Robertson (1968) asserting that a nonreduced group must contain a copy of at least one of five basic types of divisible LCA groups. Besides being of considerable intrinsic interest, the theorem will prove useful in Chapter 6.

4.23 Theorem Let $G \in \mathscr{L}$ be nonreduced. Then G contains a closed subgroup topologically isomorphic to one of the following: (a) R, (b) a nontrivial quotient of $\hat{Q}$ by a closed subgroup, (c) Q, (d) $Z(p^\infty)$, or (e) F_p for some $p \in \mathscr{P}$.

Proof: By P.29 either (a) holds or else G contains a compact open subgroup. In the latter case $c(G)$ is compact, so by 4.10 we have $c(G) \in \mathscr{S}^*(\hat{Q})$. Since $\hat{Q}$ is compact, we are led to (b) unless $c(G) = \{0\}$.

We may now assume that G is totally disconnected. By hypothesis G contains a nontrivial divisible subgroup D. We assume in this paragraph that D contains a noncompact element d. Consider the mapping $f : Z \to D$

defined by $f(n) = nd$ for each $n \in Z$. Extend f to $\bar{f} \in \mathrm{Hom}(Q,D)$ [P.9(c)]. Since d has infinite order we find that $\bar{f}$ is one-one, so that $\bar{f}(Q)$ and Q are algebraically isomorphic. We will then have case (c) if we can show that $\bar{f}(Q)$ is a discrete subgroup of G. But since $b(G)$ is open in G [P.17(c)] this will be done if we can show that $\bar{f}(Q) \cap b(G) = \{0\}$. To this end pick $y \in \bar{f}(Q) \cap b(G)$ and write $y = \bar{f}(m/n)$ for some $m/n \in Q$. Then $md = f(m) = n\bar{f}(m/n) = ny \in b(G)$. But $md \notin b(G)$ unless $m = 0$. Thus $m = 0$, whence $y = 0$, as desired. Hence $\bar{f}(Q) \cong Q$ and we have case (c).

Finally, we are left with the case that G is totally disconnected and contains a nontrivial divisible subgroup D with $D \subseteq b(G)$. Then $\bar{D} \subseteq b(G)$, so $(\bar{D})$ is totally disconnected [P.22(g)]. It follows from 3.5 that $\bar{D}$ is a topological torsion group. But then by 3.13, $\bar{D}$ is a local direct product of topological p-groups, each one of which (being a quotient of $\bar{D}$) is densely divisible. An appeal to 4.22 now leads us to case (d) or (e) and completes the proof. ∎

Having seen some applications of sufficiency classes, let us conclude by considering the question: When is a class $\mathscr{C}$ of LCA groups a sufficiency class? Some conditions are obviously necessary:

4.24 *Observation* Let $\mathscr{C}$ be a class of LCA groups. Suppose that $\mathscr{C} = \mathscr{S}(H)$ for some $H \in \mathscr{L}$. Then we have:

(a) If $G \in \mathscr{C}$ then $\mathscr{S}(G) \subseteq \mathscr{C}$. In particular, each closed subgroup of G belongs to $\mathscr{C}$.

(b) $\mathscr{C}$ is closed under the formation of finite direct products, and, more generally, under the formation of local direct products with respect to compact open subgroups.

Two frequently encountered classes of LCA groups having these two properties are the class of totally disconnected groups and the class of totally disconnected torsion-free groups. It turns out that both these classes are actually sufficiency classes (see 4.25). But we do not know in general whether conditions (a) and (b) of 4.24 on a class $\mathscr{C}$ are enough (we hesitate to use the word "sufficient") to ensure that $\mathscr{C} = \mathscr{S}(H)$ for some $H \in \mathscr{L}$. For example, is the class of all reduced LCA groups a sufficiency class?

Miscellanea

4.25 (Armacost 1971b) (a) $\mathscr{S}(Q/Z)$ is the class of all totally disconnected LCA groups.

(b) Let D be the minimal divisible extension of $\Pi_{p \in \mathscr{P}} J_p$. Then $\mathscr{S}(D)$ is the class of all totally disconnected torsion-free LCA groups. [Note that D may be alternatively described as $LP_{p \in \mathscr{P}}(F_p : J_p)$. This follows from P.31(c) and the fact that $q J_p = J_p$ if q is a prime different from p; or see §25.32 of [HR].]

4.26 The class of topological p-groups (for a fixed prime p) is not a sufficiency class. The same is true of the class of topological torsion groups.

4.27 (Armacost 1971b) Let $G \in \mathscr{L}$. Then $G = b(G)$ iff the subgroups of the form $Z(p^n)$ and J_p (where n ranges over Z^{+0} and p ranges over $\mathscr{P}$) generate a dense subgroup of G. [This can be deduced from 4.5 and 4.26(a).]

4.28 Let G and H be in $\mathscr{L}$ and let G^* and H^* be their respective minimal divisible extensions. Then if $G \in \mathscr{S}(H)$ we have $G^* \in \mathscr{S}(H^*)$ but not conversely. [Assume $G \in \mathscr{S}(H)$ and pick $x^* \neq 0$ in G^*. Now G is essential in G^* (P.8 and P.31(c)), so there exists $n \in Z$ such that $nx^* \in G$ and $nx^* \neq 0$. Pick $f \in \text{Hom}(G,H)$ such that $f(nx^*) \neq 0$. Extend f to $\bar{f} \in \text{Hom}(G^*,H^*)$ and show that $\bar{f}(x^*) \neq 0$.]

4.29 (Armacost 1971b) A class $\mathscr{C}$ of LCA groups which satisfies (a) of 4.24 and which is closed under the formation of finite direct products need not be a sufficiency class. [Let $\mathscr{C}$ be the class of LCA torsion groups of bounded order. To show that (a) holds it is perhaps quickest to recall that $G \in \mathscr{L}$ has bounded order iff $\hat{G}$ does also (cf. the proof of 3.16) and then use 4.5.]

4.30 (Armacost 1971b) If two groups determine the same sufficiency class then they must have at least one nontrivial closed subgroup in common. More precisely, if $\mathscr{S}(G_1) = \mathscr{S}(G_2)$ for G_1 and G_2 in $\mathscr{L}$, then G_i contains a closed subgroup F_i for $i = 1, 2$ such that $F_1 \cong F_2 \cong F$, where F is either Z, $Z(p)$, or J_p for some $p \in \mathscr{P}$.

4.31 Let G and H be in $\mathscr{L}$. The following are equivalent:
(a) Every $\gamma \in \hat{G}$ has the form $\gamma = \eta \circ \phi$ for some $\eta \in \hat{H}$ and $\phi \in \text{Hom}(G,H)$.
(b) $\hat{G}$ is the union of its $\hat{H}$-subgroups, i.e., $\hat{G}$ equals its $\hat{H}$-constituent.
[Property (a) can be expressed loosely by saying that each $\gamma \in \hat{G}$ can be "pulled through H." The important case in which $H = R$ (of which this is but a straightforward generalization) is due to Dixmier (1957) (see also

§24.43 of [HR]) and is useful in the study of arcwise connectedness (see 8.21). For a proof, note (P.23) that $\eta \circ \phi = \phi^*(\eta)$ for $\eta \in \hat{H}$ and $\phi \in \text{Hom}(G,H)$.]

4.32 (Cf. 4.12 and 4.13.) Let $G \in \mathscr{L}$. The H-constituent of G equals the H-component of G for $H = Z(p^\infty)$, but not in general for $H = \hat{Q}$ or T. (For $H = \hat{Q}$ or T use $G = H \times H$. Either argue directly or observe that $\hat{H} \times \hat{H}$ has one-one characters and that none of these can be "pulled through $\hat{H}$"; then use 4.31.)

4.33 (Armacost 1971b; cf. 4.20.) Let $p \in \mathscr{P}$ be fixed. A topological p-group G is densely divisible iff G/K is a weak direct product of copies of $Z(p^\infty)$ for each compact open subgroup K of G.

4.34 (Khan 1973a) Let $G \in \mathscr{L}$ be σ-compact and divisible. Then $\hat{G}$ is divisible iff G is torsion-free. [Call a group G *power-proper* iff the maps $f_n \in \text{Hom}(G,G)$ defined by $f_n(x) = nx$ for each $x \in G$ are proper. (The term is of course more appropriate when multiplicative notation for groups is used, but we retain it.) It follows from P.23(d) that if G is power-proper so is $\hat{G}$. Now let G be σ-compact, divisible, and torsion-free. Then G (and hence $\hat{G}$) is power-proper (P.30(b)) and $\hat{G}$ is densely divisible (4.15). Now use 4.16(a).]

4.35 [Based in part on Bruner (1972)] Let G and H be in $\mathscr{L}$. The H-*radical* of G [written $\mathscr{R}(G:H)$] is defined as $\{x \in G : f(x) = 0$ for each $f \in \text{Hom}(G,H)\}$. Then $\mathscr{R}(G:H)$ is a closed subgroup of G whose annihilator in $\hat{G}$ is the closure of the $\hat{H}$-component of $\hat{G}$. (This generalizes 4.5.) Furthermore, $\mathscr{R}(G:H)$ is the smallest closed subgroup K of G such that $G/K \in \mathscr{S}(H)$. For any $G \in \mathscr{L}$ we have:

    (a) $\mathscr{R}(G:R) = b(G)$.

    (b) $\mathscr{R}(G:Q/Z) = c(G)$.

    (c) $\mathscr{R}(G:Q) = b(G) + c(G)$. [By P.17(c) we see that $\mathscr{R}(G:Q)$ is open in G. Dually, the $\hat{Q}$-component of any LCA group is relatively compact.]

    (d) $\mathscr{R}(G:\hat{Q})$ plays the role of the torsion subgroup in abelian group theory in this sense: $\mathscr{R}(G:\hat{Q})$ is the smallest closed subgroup K of G such that G/K is torsion-free. We have $\overline{t(G)} \subseteq \mathscr{R}(G:\hat{Q})$, but the inclusion may be proper. [Let A be a reduced group in $\mathscr{L}_d$ satisfying $\cap_{n=1}^\infty nA \neq \{0\}$ (see P.9(h)) and set $G = \hat{A}$. Then $\overline{t(G)} \subsetneqq G = \mathscr{R}(G:\hat{Q})$.] Finally, $\mathscr{R}(G:\hat{Q})$ need not be a subset of G!. ($G = T$; cf. 3.4.)

We note finally that the paper of Bruner (1972) contains a good deal of information and several queries about radicals. We mention one such question, which may be of interest. One can show without much trouble that the H-radicals of (a) through (d) are idempotent in the sense that $\mathscr{R}(\mathscr{R}(G:H):H) = \mathscr{R}(G:H)$ for all $G \in \mathscr{L}$. It seems to be a difficult problem to characterize those $H \in \mathscr{L}$ having this property. An example of $H \in \mathscr{L}$ without this property is $H = \Pi_{n=2}^{\infty} Z(n)$. It turns out that $\mathscr{R}(G:H) = \cap_{n=1}^{\infty} \overline{nG}$ for each $G \in \mathscr{L}$. Hence if G is a discrete group as in P.9(h) we have $\mathscr{R}(\mathscr{R}(G:H):H) \subsetneqq \mathscr{R}(G:H)$.

4.36 The reader will have observed that we have not determined any sufficiency class $\mathscr{S}(H)$ where H is not divisible, i.e., where Lemma 4.6 does not apply. For example, we do not have an adequate description of $\mathscr{S}(J_p)$. We can, however, determine $\mathscr{S}(Z)$. [The determination of $\mathscr{S}(Z)$ is contained implicitly in Khan (1973a).] We have: $G \in \mathscr{L}$ belongs to $\mathscr{S}(Z)$ iff G is discrete and is a subgroup of a (full) direct product of copies of Z. [It is clear from P.29 that if $G \in \mathscr{S}(Z)$ then G is discrete. Now for each $x \neq 0$ in G find $f_x \in \text{Hom}(G,Z)$ such that $f_x(x) \neq 0$. Then set $H = \Pi Z_x$, where $Z_x \cong Z$ for each $x \neq 0$ in G, and define $\phi : G \to H$ by $\phi(g) = (f_x(g))$ for each $g \in G$. Then ϕ is a monomorphism, so G has the form mentioned. The converse is evident.]

4.37 Let $G \in \mathscr{L}$. If G is *self-dual* (i.e., $G \cong \hat{G}$) then (trivially) we have $G \in \mathscr{S}(\hat{G})$ and $\hat{G} \in \mathscr{S}(G)$, but the converse fails. [It may still be possible to shed some light on the difficult problem of classifying the self-dual LCA groups by studying those $G \in \mathscr{L}$ satisfying $G \in \mathscr{S}(\hat{G})$, although the author is at present doubtful. We digress at this point to discuss briefly what seems to be known about the self-dual LCA groups. Familiar examples are the finite groups, R, F_p, and of course groups of the form $G \times \hat{G}$. Other examples are given in §25.34 of [HR]. An important step forward was taken by Rajagopalan and Soundararajan (1969), who give a complete classification of the self-dual torsion-free $G \in \mathscr{L}$ which satisfy certain additional topological conditions, such as metrizability or σ-compactness. [This paper corrects an overly hasty announcement in their earlier paper (1967) where such additional conditions were omitted.] Shortly afterwards, Corwin (1970) initiated a new and interesting approach to the problem of classifying the self-dual LCA groups. In the decade since the appearance of these two papers, however, no new progress appears to have been made, and a definitive solution of the problem awaits further insights.]

4.38 (Armacost 1971a) Let $G \in \mathscr{L}$. Then $\text{Hom}(G,\hat{G}) = \{0\}$ iff (a) G is densely divisible, (b) $b(G) = G$, and (c) $b(\hat{G})$ is reduced. [The author, having temporarily given up on the problem of the self-dual groups, decided to console himself with an investigation of the "anti-self-dual" groups, i.e., those groups $G \in \mathscr{L}$ such that both $\text{Hom}(G,\hat{G})$ and $\text{Hom}(\hat{G},G)$ are trivial. He found, by using the characterization above, that no such groups (besides the trivial group) exist (Armacost 1971a). If nothing else, this result shows at least that G and $\hat{G}$ are never totally "incompatible." A more direct proof of the nonexistence of nontrivial anti-self-dual groups may be found in Olubummo and Rajagopalan (1976).]

4.39 *Envoi* There are several other topics, either closely or tangentially related to the subject of this chapter, which we do not explore. We commend to the reader the paper of Khan (1973a) who studies (among several other things) conditions under which there exist sufficiently many *proper* homomorphisms from one LCA group to another. Several authors have studied the topological group structure of $\text{Hom}(G,H)$. We mention Preston (1956), who gives an interesting analogue of the duality theorem (valid for a proper subclass of $\mathscr{L}$) using as "character group" the group $\text{Hom}(G,Q/Z)$ in place of $\hat{G}$. (Preston's argument appears to apply only to topological torsion groups, and not to the totally disconnected groups as claimed.) We also mention Moskowitz (1967), who gives conditions under which $\text{Hom}(G,H)$ becomes an LCA group. A more specialized topic is the study of the group of topological automorphisms of an LCA group. Besides §26 of [HR], we refer the reader to the following papers: Braconnier (1948, Chapitre IV), Keesling (1972), Levin (1971), and Robertson (1967).

5

H-Dense Groups

"All these were honored in their generations." [*Ecclesiasticus* 44:7]

If an LCA group G contains an element x which generates a dense subgroup of G, then G is called monothetic (P.25), and we refer to the element x as a *generator* of G. A basic and important fact about monothetic LCA groups is that with the exception of Z, they are all compact (P.25). In this chapter we shall discuss the monothetic groups in some detail. We shall also investigate two related classes of groups having, in some respects, remarkably similar properties.

5.1 Definition Let $H \in \mathscr{L}$ be fixed. A group $G \in \mathscr{L}$ is said to be *H-dense* iff there exists $f \in \text{Hom}(H,G)$ such that $f(H)$ is a dense subgroup of G, i.e., (in the language of 4.2) iff G has a dense H-subgroup.

Thus the monothetic groups are the Z-dense groups. The R-dense groups, that is, those containing a dense one-parameter subgroup (see after 4.7), are called *solenoidal*. Finally, if $p \in \mathscr{P}$ is fixed, the $Z(p^\infty)$-dense groups are referred to, perhaps infelicitously (Armacost and Armacost 1972a), as *p-thetic*. We shall take up these three types in order and close by stating a result (whose proof we postpone until Chapter 10) which links these three types of groups together, to the exclusion of other H-dense groups, by a common property. Since the monothetic groups have been studied for some time, and since they are of particular importance in other areas of mathematics, it is not surprising that most of our results will deal with them. (See [HR] for historical notes on monothetic groups.) First, a simple but useful statement.

5.2 Proposition Let $H \in \mathscr{L}$ be fixed and let $G \in \mathscr{L}$.

(a) If G is H-dense and $G' \in \mathscr{L}$ is a continuous homomorphic image of G, then G' is H-dense.

(b) G is H-dense iff there exists a continuous monomorphism from $\hat{G}$ into $\hat{H}$.

Proof: (a) is obvious, while (b) follows from P.23(b). ∎

The following immediate consequence of part (b) [first found by Halmos and Samelson (1942) and also proved shortly afterwards by Anzai and Kakutani (1943) and Eckmann (1944)] is, despite its apparent simplicity in this setting, of fundamental importance in the study of monothetic groups.

5.3 Corollary A compact group $G \in \mathscr{L}$ is monothetic iff $\hat{G}$ is (isomorphic to) a subgroup of T_d. ∎

To be able to use 5.3 it will be helpful to know how to recognize a subgroup of T_d. The following fact is well known (see A.14 and §25.13 of [HR]).

5.4 Lemma We have $T_d \cong Q^{c*} \times \Pi^*_{p \in \mathscr{P}} Z(p^\infty)$. Hence a group $A \in \mathscr{A}$ is isomorphic to a subgroup of T_d iff $r_0(A) \leq c$ and $r_p(A) \leq 1$ for all $p \in \mathscr{P}$.

Proof: Clearly $t(T_d) \cong Q/Z \cong \Pi^*_{p \in \mathscr{P}} Z(p^\infty)$ [P.9(e)]. Since T_d is divisible, so is $t(T_d)$ whence [P.9(d)] we have $T_d \cong D \times t(T_d)$, where $D \cong T_d / t(T_d)$ is torsion-free, divisible, and has cardinality (hence rank) equal to c. Hence $D \cong Q^{c*}$ by P.9(e), which proves the first statement. The second statement follows immediately from P.9(g). ∎

Before giving examples of monothetic groups, we mention some notation. If $G \in \mathscr{L}$ we let $e(G)$ denote the set of generators of G, so that G is monothetic iff $e(G)$ is not empty. The complement of $e(G)$, consisting of "nongenerators," will be denoted by $n(G)$. The facts appearing in the next section are well known (and may be found described in [HR]), but we include them here for completeness.

5.5 Examples and Observations (a) T is of course monothetic. It is evident that $n(T) = t(T)$. In fact, this property characterizes T among the infinite monothetic groups (see 5.14).

(b) A torus T^m is monothetic iff $m \leq c$. (It may seem surprising that so large a group as T^c could have a dense cyclic subgroup, but all we need do is observe that Z^{m*} has rank m and apply 5.3 and 5.4.)

(c) More generally, if $G \in \mathscr{L}$ is compact and connected, then G is monothetic iff $w(G) \leq c$. [By P.19(e) we have $|\hat{G}| = w(\hat{G}) = w(G)$. If then $w(G) > c$, $\hat{G}$ could not be a subgroup of T_d, so G could not (5.3) be monothetic. Conversely, if $w(G) \leq c$, we have $|\hat{G}| \leq c$, and since $\hat{G}$ is torsion-free (P.28(b)) we see that $r_0(\hat{G}) \leq c$ and $r_p(\hat{G}) = 0$ for all $p \in \mathscr{P}$, so by 5.3 and 5.4, G is monothetic.] In particular, the groups $\hat{Q}$ and $(R_d\hat{)}$ are monothetic.

(d) Since each group $Z(p^\infty)$ is a subgroup of T_d, we find that the groups J_p are monothetic. In fact, since each closed subgroup of J_p has the form L_k (see P.18) it follows that $e(J_p)$ consists of all those $\bar{x} = (x_0, x_1, \ldots) \in J_p$ with $x_0 \neq 0$. We observe then, that $n(G) = L_1$, so $n(G)$ is an open (hence closed) subgroup of G. (See 5.10 for a converse.) Hence $e(J_p)$ is an open and closed subset of J_p.

(e) A compact totally disconnected group $G \in \mathscr{L}$ is monothetic iff $G \cong \Pi_{p \in \mathscr{P}} A_p$, where for each $p \in \mathscr{P}$ we have $A_p \cong J_p$ or $A_p \cong Z(p^n)$ for some $n \in Z^{+0}$. [For the discrete torsion group $\hat{G}$ may be written (P.6(a)) as $\Pi^*_{p \in \mathscr{P}} (\hat{G})_p$. Clearly $\hat{G}$ is isomorphic to a subgroup of T_d iff $(\hat{G})_p$ has the form $Z(p^n)$ or $Z(p^\infty)$ for each $p \in \mathscr{P}$.] Another way to say this is: A compact totally disconnected group $G \in \mathscr{L}$ is monothetic iff G is a quotient of $(Q/Z\hat{)}$ by a closed subgroup.

(f) A compact group $G \in \mathscr{L}$ is monothetic iff $w(G) \leq c$ and $G/c(G)$ is a quotient of $(Q/Z\hat{)}$ by a closed subgroup. [Assume G monothetic. Then since $\hat{G}$ is a subgroup of T_d we have $w(G) = |\hat{G}| \leq c$; moreover, $G/c(G)$ is totally disconnected and monothetic and is hence by (e) a quotient of $(Q/Z\hat{)}$ by a closed subgroup. Conversely, if $w(G) \leq c$ then $r_0(\hat{G}) \leq |\hat{G}| \leq c$; if moreover $G/c(G)$ is a quotient of $(Q/Z\hat{)}$ it follows that $t(\hat{G}) \cong (G/c(G))\hat{}$ is a subgroup of Q/Z, so $r_p(\hat{G}) \leq 1$ for each $p \in \mathscr{P}$. Thus by 5.3 and 5.4, G is monothetic.]

(g) The group $(T_d\hat{)}$ is often called a "universal compact monothetic group," since it follows immediately from 5.3 that a compact group $G \in \mathscr{L}$ is monothetic iff G is a quotient of $(T_d\hat{)}$ by a closed subgroup. [Note that $T \times (T_d\hat{)}$ is compact monothetic and has the same property, but $T \times (T_d\hat{)} \not\cong (T_d\hat{)}$.]

5.6 Remark We note in (g) above that $(T_d\hat{)}$ is just the Bohr compactification $\beta(Z)$ of Z (see P.32). One may easily prove in general that for any $H \in \mathscr{L}$ a compact group $G \in \mathscr{L}$ is H-dense iff G is a quotient of $\beta(H)$ by a closed subgroup. Hence we may refer to $\beta(H)$ as a "universal compact H-dense group."

We now appropriate from [HR] (§25.11) a simple and helpful fact about generators.

5.7 Proposition Let $G \in \mathscr{L}$. An element $x \in G$ belongs to $e(G)$ iff $\gamma(x) \neq 1$ for each $\gamma \neq 1$ in $\hat{G}$. (Hence the generators of G may be characterized as those elements $x \in G$ which, considered as continuous characters of $\hat{G}$, are monomorphisms.)

Proof: If $x \in e(G)$ and $\gamma(x) = 1$, then $\gamma(\overline{\text{gp}(x)}) = \gamma(G) = \{1\}$, so $\gamma = 1$. Conversely, if $x \notin e(G)$ then [P.21(a)] $A(\hat{G}, \overline{\text{gp}(x)}) \neq \{1\}$, i.e., there exists $\gamma \neq 1$ in $\hat{G}$ such that $\gamma(x) = 1$. ∎

We are now in a position to answer some natural questions about generators and nongenerators. For example, $e(T)$ is dense in T [cf. 5.5(a)] while $e(J_p)$ is a proper closed subset of G [5.5(d)]. We are tempted to suppose that the density of $e(G)$ has something to do with the connectedness of G. In fact, we have the following result from Armacost (1971c) [the argument for which, we have subsequently learned, is essentially that of Corollary 4.1 of Baayen and Helmberg (1965)].

5.8 Proposition Let $G \in \mathscr{L}$ be monothetic. Then $e(G)$ is dense in G iff G is connected.

Proof: We may certainly assume that G is not discrete, whence by P.25, G is compact. Suppose that G is connected. Let $x \in e(G)$ and let $n \neq 0$ be an integer. We claim that $nx \in e(G)$ also. For suppose $\gamma(nx) = 1$ for some $\gamma \in \hat{G}$. Then $\gamma^n(x) = 1$, whence by 5.7 we have $\gamma^n = 1$. Since $\hat{G}$ is torsion-free [P.28(b)] we have $\gamma = 1$, whence by 5.7 again, $nx \in e(G)$. We now have $\text{gp}(x) \cap \{0\}^c \subseteq e(G)$. However, $\text{gp}(x)$ is dense in G, and since G is not discrete, $\text{gp}(x) \cap \{0\}^c$ must be dense in G too. Thus $e(G)$ is dense in G. Conversely, if G is not connected, let U be a proper open subgroup of G [P.27(c)]. Obviously $e(G) \cap U$ is empty, so $e(G)$ is not dense in G. ∎

It would be of interest to find an explicit description of $\overline{e(G)}$ in an arbitrary compact monothetic group. Leaving this problem to the curious reader, we now ask under what circumstances $n(G)$ is dense in G. The following result (from Armacost 1971c) shows that $n(G)$ fails to be dense in G for relatively few groups G.

5.9 Proposition Let G be compact and monothetic. If G is not totally disconnected then $n(G)$ is dense in G. If $n(G)$ is not dense in G then $n(G)$ is closed, and G is a product of finitely many of the groups A_p described in 5.5(e).

Proof: We take a paragraph to describe the generators of a group A of the form $\Pi_{p \in \mathscr{P}} A_p$ as in 5.5(e). Following Halmos and Samelson (1942) (or §25.27 of [HR]) we show that if $\bar{a} = (a_p) \in A$ then $\bar{a} \in e(A)$ iff $a_p \in e(A_p)$ for all $p \in \mathscr{P}$. Suppose first that $a_p \in e(A_p)$ for all $p \in \mathscr{P}$. Let γ_{p_i} belong to $(A_{p_i})\hat{}$ for distinct primes $p_1, \ldots, p_n$. Since $\gamma_{p_i}(A_{p_i}) \subseteq T_{p_i}$ for $i = 1, \ldots, n$ and since $t(T)$ ($\simeq Q/Z$) is the weak direct product of its subgroups $T_p(\simeq Z(p^\infty))$ [cf. 2.6 and P.6(a)] we see that if $\gamma_{p_1}(a_1) \cdots \gamma_{p_n}(a_n) = 1$ then $\gamma_{p_1}(a_1) = \cdots = \gamma_{p_n}(a_n) = 1$, so each γ_{p_i} is trivial (5.7). We then conclude from the description of $\hat{A}$ [P.19(b)] that if $\gamma \in \hat{A}$ and $\gamma(\bar{a}) = 1$ then $\gamma = 1$, whence by 5.7, $\bar{a} \in e(A)$. Conversely, it is obvious that if $a = (a_p) \in e(A)$ then $a_p \in e(A_p)$ for each $p \in \mathscr{P}$.

Now to the proof of the proposition. If G is not totally disconnected find $\gamma \in \hat{G}$ having infinite order [P.28(b)]. Then by 5.7 we have $\cup_{n=1}^\infty \ker \gamma^n \subseteq n(G)$. But from P.22(j) we see that $A(\hat{G}, \cup_{n=1}^\infty \ker \gamma^n) = \cap_{n=1}^\infty \overline{\mathrm{gp}(\gamma^n)} = \cap_{n=1}^\infty \mathrm{gp}(\gamma^n) = \{1\}$, so the subgroup $\cup_{n=1}^\infty \ker \gamma^n$ is dense in G, whence $n(G)$ is dense too. Hence if $n(G)$ is not dense in G, G is totally disconnected, and we may assume G is a group $A = \Pi_{p \in \mathscr{P}} A_p$ described in 5.5(e). If $\overline{n(A)} \subsetneqq A$ it is clear from the first paragraph and the definition of the product topology that almost all the groups A_p are trivial. We must show in this case that $n(A)$ is closed in A. Now $e(A_p)$ is open in A_p for each $p \in \mathscr{P}$ [for the case $A_p = J_p$ see 5.5(d); for $A_p = Z(p^n)$ this is trivial]. Hence if only finitely many A_p's are nonzero, it follows from the first paragraph that $e(A) = \Pi_{p \in \mathscr{P}} e(A_p)$ is open in A, so $n(A)$ is closed. [We also note for the record that it follows from the first paragraph that $n(G)$ is always open for totally disconnected G. For the converse see 5.12.] ∎

We now impose algebraic as well as topological conditions on $n(G)$ to obtain the following characterization (Armacost 1971c) of the p-adic integer groups. (Compare with 5.25.)

5.10 Corollary The following are equivalent for an infinite compact monothetic group G:

(a) $n(G)$ is a closed subgroup of G.

(b) $n(G)$ is an open subgroup of G.

(c) $G \cong J_p$ for some $p \in \mathscr{P}$.

Proof: Assume (a). Then from 5.9, G is the product of finitely many groups A_p. If more than one of the groups A_p is nonzero, it is clear we may form a generator by adding two appropriate nongenerators, so $n(G)$ is not a subgroup of G. Thus $G \cong A_p$ for some p, and since G is infinite we have $G \cong J_p$, i.e., (a) $\Rightarrow$ (c). The implication (c) $\Rightarrow$ (b) is given in 5.5(d), while (b) $\Rightarrow$ (a) is trivial. ∎

We have seen from the proof of 5.9 that $n(G)$ is an open subset of the monothetic group G whenever G is totally disconnected. [This also follows from P.28(c) and 5.7.] What is not so obvious is that the converse is true as well. This is an immediate consequence of our next result, which we insert here for this reason and also because it is a rather deep result of independent interest. By way of introduction to it, we observe that if G and H in $\mathscr{L}$ are monothetic and $f \in \operatorname{Hom}(G,H)$ is surjective then $f(x) \in e(H)$ whenever $x \in e(G)$. But it is not clear that we can go the other way; that is, if $y \in e(H)$ does there exist $x \in e(G)$ such that $f(x) = y$? In a very important case, we can answer this in the affirmative, as the following result of Baayen and Helmberg (1965) shows. Our proof adopts a simplification due to Kuipers and Niederreiter (1974), but is a bit different. (For a generalization see 5.30.)

5.11 *Theorem* Let G be a compact monothetic group and let $\gamma \in \hat{G}$ be surjective. For each $t \in e(T)$ there exists $x \in e(G)$ such that $\gamma(x) = t$.

Proof: Let us first make an observation about T_d. We claim that if a and b belong to T_d and have infinite order, then there is an automorphism f of T_d such that $f(a) = b$. One way to see this is as follows. Let D be a subgroup of T_d containing $\operatorname{gp}(a)$ such that $D \simeq (\operatorname{gp}(a))^*$, the minimal divisible extension of $\operatorname{gp}(a)$ [P.9(g)]. Evidently $D \simeq Q$, and by P.9(d) we have $T_d = D \dotplus K$ for some subgroup K of T_d. It is clear by a rank argument and 5.4 that $K \simeq T_d$. Hence we can find an isomorphism ϕ carrying T_d onto $Q \times T_d$ such that $\phi(a) = (r,0)$ for some $r \neq 0$ in Q. In the same way we can find an isomorphism ψ from T_d onto $Q \times T_d$ such that $\psi(b) = (s,0)$ for some $s \neq 0$ in Q. Now define an automorphism μ of $Q \times T_d$ by $\mu(c,d) = (sc/r,d)$ for each $c \in Q$ and $d \in T_d$. Then $f = \psi^{-1} \circ \mu \circ \phi$ is an automorphism of T_d such that $f(a) = b$.

Now we take up the proof of the theorem. It follows immediately from

5.5(g) that the result will be proved if we prove it for the case $G = (T_d)\hat{}$. Accordingly, we set $G = (T_d)\hat{}$ and seek some $x \in e((T_d)\hat{})$ such that $\gamma(x) = t$. By the duality theorem we may think of γ as an element of T_d. Clearly $\gamma \in T_d$ has infinite order, since γ is surjective as a map; likewise for t, since $t \in e(T)$. Thus we seek some character x on T_d such that x is a generator of $(T_d)\hat{}$ and such that $x(\gamma) = t$. To ask that x be a generator of $(T_d)\hat{}$ is to ask, in light of 5.7, that x be a monomorphism from T_d to T_d. If in the first paragraph we take $a = \gamma$ and $b = t$, we see that the automorphism f may be used for x. ∎

The following promised consequence of Theorem 5.11 is also from Baayen and Helmberg (1965). [The result also appears, with a very different proof, in a paper (Armacost 1971c) of the author, who was unaware at the time of the work of Baayen and Helmberg (cf. Armacost 1973).]

5.12 Corollary Let $G \in \mathscr{L}$ be compact and monothetic. Then $n(G)$ is open in G iff G is totally disconnected.

Proof: We have already shown that if G is totally disconnected then $n(G)$ is open in G. Conversely, assume that $n(G)$ is open and hence that $e(G)$ is compact. If G is not totally disconnected, find $\gamma \in \hat{G}$ such that $\gamma(c(G)) \neq \{1\}$. Then γ is surjective [P.28(a)], whence by 5.11, $\gamma(e(G)) = e(T)$. But then $e(T)$ is compact, contradicting the fact [5.5(a)] that $e(T)$ is the complement of $t(T)$. ∎

The following consequence of 5.11 may be well known, although the author has no reference at hand.

5.13 Corollary Let G be a compact infinite monothetic group. Then $|e(G)| \geqq c$.

Proof: First assume that G is not totally disconnected. Then there exists a surjective $\gamma \in \hat{G}$. By 5.11, $\gamma(e(G)) = e(T)$. But clearly $|e(T)| = c$, whence $|e(G)| \geqq c$. If on the other hand G is totally disconnected, we may take G to be $A = \Pi_{p \in \mathscr{P}} A_p$ as in 5.5(e). We showed in the first paragraph of the proof of Proposition 5.9 that $e(A) = \Pi_{p \in \mathscr{P}} e(A_p)$. If then $A_p = J_p$ for some p we have $|e(A)| = c$, since $|e(J_p)| = c$ [see 5.5(d)]. If none of the A_p's is J_p then (since G is assumed infinite) infinitely many of the A_p's are nontrivial. Since $|e(Z(p^n))| \geqq 2$ for all primes greater than 2 we again have $|e(A)| = c$. ∎

We conclude our investigation of monothetic groups by taking up a point mentioned in 5.5(a). The circle T is in a sense quite egalitarian, in that each of its elements having an outside chance of generating T (i.e., having infinite order) actually does generate T. We now show that T is the only infinite compact monothetic group with this property. (This is perhaps not surprising in that Z is the "smallest" infinite subgroup of T_d.) The result is from a paper of the author (Armacost 1971c).

5.14 Proposition Let G be an infinite compact monothetic group. If each element of G having infinite order is a generator of G, then $G \cong T$.

Proof: The statement may be proved by reducing it to a problem in abelian group theory, but we find the following argument far more amusing. A glance at 5.5(e) shows that G cannot be totally disconnected. Now let H be any proper closed subgroup of G. Since $H \subseteq n(G)$ it follows that H is a torsion group. In particular, H cannot be connected (cf. 3.5). Thus $c(G) = G$, and it follows from 4.8 that there exists $f \neq 0$ in Hom(R, G). Since $\overline{f(R)}$ is a closed connected subgroup of G, we see that $\overline{f(R)}$ cannot be proper, so $G = \overline{f(R)}$ (in other words, G is solenoidal). Now pick any $\gamma \neq 1$ in $\hat{G}$ and define $\phi \in$ Hom(R, T) by $\phi = \gamma \circ f$. Since $f(R)$ is connected and T has no proper connected subgroups, we see that ϕ is surjective. Therefore ϕ cannot be one-one, so there exists $r \neq 0$ in R such that $\gamma(f(r)) = 1$. Hence (5.7) $f(r) \in n(G)$, so $f(r)$ has finite order. Thus f is not one-one, so by P.23(b), $f^* \in$ Hom$(\hat{G}, \hat{R})$ does not have dense image. It follows from P.18(b) and the fact that $\hat{R} \cong R$ that $\overline{f^*(\hat{G})} \cong Z$. But then $f^*(\hat{G}) \cong Z$, and since f^* is one-one [by P.23(b) and the fact that f has dense image] we get $\hat{G} \cong Z$, whence $G \cong T$, as desired. ∎

We now turn to a brief discussion of the solenoidal groups. These groups are much like the monothetic groups in the sense of the following proposition. [This is a special case of §9.1 of [HR]; our proof, however, is modeled on an argument in Ross (1965) used to give an alternative proof of Lemma 2.1 of Hewitt (1963).]

5.15 Proposition Let $G \in \mathscr{L}$ be solenoidal. Then either $G \cong R$ or else G is compact.

Proof: Let $f \in$ Hom(R, G) have dense image. Then $f^* \in$ Hom$(\hat{G}, \hat{R})$ is one-one [P.23(b)]. Since $\hat{R}$ ($\cong R$) has no proper connected subgroups, we see from P.29 applied to $\hat{G}$ that either $\hat{G} \cong R$ or else $\hat{G}$ has a compact open subgroup H. In the latter case, we have $f^*(H) \subseteq b(\hat{R}) = \{0\}$, so $H = \{0\}$ and $\hat{G}$ is discrete. The result now follows by duality. ∎

In the proof of 5.14 it turned out that the compact connected monothetic group G was solenoidal. This was no accident. We have the following characterization of the compact solenoidal groups (cf. §25.18 of [HR]).

5.16 Proposition The following are equivalent for a compact $G \in \mathscr{L}$:
   (a) G is solenoidal.
   (b) $\hat{G}$ is (isomorphic to) a subgroup of R_d.
   (c) G is connected and $w(G) \le \mathfrak{c}$.
   (d) G is connected and monothetic.

Proof: The equivalence of (a) and (b) is a special case of 5.2(b). From P.28(b) and P.19(e) we have (b) ⇒ (c), while from 5.5(c) we get (c) ⇒ (d). Finally, assume (d). Then $\hat{G}$ is a torsion-free subgroup of T_d, whence, by taking minimal divisible extensions, $\hat{G}$ is (isomorphic to) a subgroup of $Q^{\mathfrak{c}*}$. Since $Q^{\mathfrak{c}*}$ is algebraically isomorphic to R [P.9(e)] it follows from 5.2(b) that G is R-dense, i.e., (d) ⇒ (a). ∎

5.17 Examples Besides the familiar solenoidal groups R, T, and $\hat{Q}$ there are two types worthy of explicit mention.
   (a) Let B be the subgroup of $R \times J_p$ consisting of all integral multiples of $(1, \bar{u})$, where $\bar{u} = (1, 0, 0, \dots) \in J_p$. Then B is a discrete, hence closed, subgroup of $R \times J_p$. The *p-adic solenoid* $\sum_p$ is defined as $(R \times J_p)/B$. It turns out (§10.13 of [HR]) that $\sum_p$ is indeed a compact solenoidal group. In fact, $(\sum_p)\hat{}$ is just the subgroup Q_p of Q consisting of all rational numbers with denominator a power of p (see §25.3 of [HR]). [Using the generalization $J_{\bar{a}}$ of J_p one can define analogously the groups $\sum_{\bar{a}}$, which again are solenoidal (§§10.12 and 10.13 of [HR]). In fact, if $\bar{a} = (2, 3, 4, \dots)$ then $\sum_{\bar{a}}$ turns out (§25.4 of [HR]) to be our familiar friend $\hat{Q}$. We shall, however, have no need for this alternative description of $\hat{Q}$.]
   (b) The group $(\hat{Q})^{\mathfrak{c}}$ is by 5.16 a solenoidal group. It derives its importance from the fact that $(\hat{Q})^{\mathfrak{c}} \cong \beta(R)$ and is hence a universal compact solenoidal group in the sense of 5.6.

Before leaving the solenoidal groups let us investigate the implication (a) ⇒ (d) in Proposition 5.16 a bit more closely. A compact solenoidal group G must have generators [indeed $e(G)$ is dense in G by 5.8], but what is the connection between $e(G)$ and the dense one-parameter subgroups of G? One obvious question to ask is this: If $f \in \text{Hom}(R,G)$ has dense image, can it happen that $f(r) \in e(G)$ for all $r \neq 0$ in R? It is easy to answer this in the negative if G is nontrivial. Indeed, let $\gamma \neq 1$ be in $\hat{G}$. Since $\gamma(f(R))$ is a nontrivial connected subgroup of T we have $\gamma(f(R)) = T$. Hence there exists $r \in R$ such that $\gamma(f(r)) = -1$, whence $\gamma(f(2r)) = 1$, so by 5.7, $f(2r) \notin e(G)$. However, if G is "small" enough then "most" $r \in R$ satisfy $f(r) \in e(G)$, as the following result, stated in Anzai and Kakutani (1943), shows.

5.18 *Proposition* Let G be a compact metrizable solenoidal group. If $f \in \text{Hom}(R, G)$ has dense image then the set $N = \{r \in R : f(r) \in n(G)\}$ is countably infinite.

Proof: Pick any $\gamma \neq 1$ in $\hat{G}$. Then $\gamma \circ f \in \hat{R}$ is nonzero, so $B_\gamma = \ker(\gamma \circ f)$ is a proper closed subgroup of R. Hence [P.18(b)] we have $B_\gamma \cong Z$, whence of course $|B_\gamma| = \aleph_0$. By 5.7 we see that N is the union of the sets B_γ over all $\gamma \neq 1$ in $\hat{G}$, and since $|\hat{G}| = \aleph_0$ (P.33) we have $|N| = \aleph_0$, as desired. ∎

If the metrizability condition on G is dropped, the situation may change radically: An entire dense one-parameter subgroup of G may fail to meet a single generator, as the following example, adapted from Anzai and Kakutani (1943), shows. (Compare with 5.28.)

5.19 *Example* There exists a compact solenoidal group G and $f \in \text{Hom}(R, G)$ having dense image such that $f(r) \in n(G)$ for all $r \in R$. We construct G as follows. For each $s \in R$ let T_s be a copy of T and set $P = \Pi_{s \in R} T_s$. Define $f: R \to P$ by letting f of any $r \in R$ be the element in P whose sth coordinate is $\exp(isr)$. It is clear that f is a continuous monomorphism from R into P. Letting G be the closure of $f(R)$ in P, we see that G is a compact solenoidal group (in fact, G is just $\beta(R)$: see §26.12 of [HR]). We show that $f(r) \in n(G)$ for each $r \in R$. Let π_s be the projection of P onto T_s and let π_s' be the restriction of π_s to G. Clearly $\pi_s' \in \hat{G}$ for each $s \in R$ and $s \neq 0 \Rightarrow \pi_s' \neq 1$. Now pick any $r \in R$. We have $\pi_s'(f(r)) = \exp(isr)$ for all $s \in R$. Setting $s = 1$ in the trivial case $r = 0$ or $s = (2\pi)/r$ otherwise, we have $\pi_s'(f(r)) = 1$. Since π_s' is a nontrivial element of $\hat{G}$ we conclude from 5.7 that $f(r) \in n(G)$. ∎

We now take up the p-thetic groups. These groups are similar to the monothetic and solenoidal groups in a sense made precise in the next proposition. [This and most of the remaining results on p-thetic groups are derived from Armacost and Armacost (1972a). The argument in 5.20 is based on a generalization of a result appearing in Rajagopalan (1968b) and Rickert (1967); see 10.1 below.]

5.20 Proposition Let $G \in \mathscr{L}$ be p-thetic for some $p \in \mathscr{P}$. Then either $G \cong Z(p^\infty)$ or else G is compact.

Proof: Let $f \in \mathrm{Hom}(Z(p^\infty), G)$ have dense image. By duality, we wish to show that either $\hat{G} \cong J_p$ or else $\hat{G}$ is discrete. Now the adjoint map $f^* \in \mathrm{Hom}(\hat{G}, J_p)$ is one-one [P.23(b)]. Since $c(J_p) = \{0\}$ and $f^*(c(\hat{G})) \subseteq c(J_p)$, we conclude that $\hat{G}$ is totally disconnected and hence (P.29) has a compact open subgroup H. If H is trivial, then $\hat{G}$ is discrete. So we assume that H is not trivial, whence $f^*(H)$ is a nontrivial compact subgroup of J_p. By P.18(d), $f^*(H)$ is then an open subgroup of J_p. Since H is compact, the restriction of f^* to H is an open mapping from H onto $f^*(H)$, whence by P.30(a), f^* is an open mapping. Therefore f^* is a topological isomorphism from $\hat{G}$ onto $f^*(\hat{G})$, and since by P.18(d) we have $f^*(\hat{G}) \cong J_p$, we conclude that $\hat{G} \cong J_p$. ∎

Before presenting examples of p-thetic groups we narrow our search by the following simple fact. It is helpful to recall that J_p is a torsion-free group of cardinality $\mathfrak{c}$. Since $r(J_p)$ is clearly infinite we also have $r(J_p) = \mathfrak{c}$ (see P.5).

5.21 Proposition A compact group $G \in \mathscr{L}$ is p-thetic for a given $p \in \mathscr{P}$ iff $\hat{G}$ is (isomorphic to) a subgroup of $(J_p)_d$. Hence a compact p-thetic group is solenoidal and therefore connected and monothetic.

Proof: The first statement is a special case of 5.2(b). Now for any $p \in \mathscr{P}$, $(J_p)_d$ is a subgroup of R_d, since by P.9(g) the minimal divisible extension of $(J_p)_d$ is Q^{c*}, which is isomorphic to R_d. The proof is now completed by appealing to 5.16. ∎

5.22 Examples (a) It follows from 5.5(b) and 5.21 that $T^{\mathfrak{m}}$ cannot be p-thetic for any prime p if $\mathfrak{m} > \mathfrak{c}$. On the other hand, if $\mathfrak{m} \leq \mathfrak{c}$ then $T^{\mathfrak{m}}$ is p-thetic for all primes p. Indeed, since $r(J_p) = \mathfrak{c}$ it is clear that for such $\mathfrak{m}$, $(J_p)_d$ contains a subgroup isomorphic to $Z^{\mathfrak{m}*}$, so $T^{\mathfrak{m}}$ is p-thetic by 5.21. (There are groups not of this form which are p-thetic for all primes p; see 5.34.)

(b) Let $p \in \mathcal{P}$ be fixed. Then the p-adic solenoid $\sum_p$ is not p-thetic, but is q-thetic for each $q \in \mathcal{P}$ different from p. To see that $\sum_p$ is not p-thetic, recall [5.17(a)] that $(\sum_p)\hat{}$ is the subgroup Q_p of Q consisting of all rationals having denominator a power of p. Since $\cap_{n=1}^{\infty} p^n J_p = \{0\}$ it is clear that Q_p cannot be a subgroup of $(J_p)_d$, so $\sum_p$ is not p-thetic. However, if $q \neq p$ we have $qJ_p = J_p$ [cf. P.22(f)], whence $(J_p)_d$ contains a copy of Q_q. Therefore $\sum_p$ is q-thetic.

(c) There are of course compact solenoidal groups that are p-thetic for no $p \in \mathcal{P}$. Indeed, since $t(G)$ is dense in G if G is p-thetic, it is clear that $\hat{Q}$ is an example of this. But there exist compact solenoids having dense torsion subgroup which are p-thetic for no prime p. Let $G = \Pi_{p \in \mathcal{P}} \sum_p$. From 5.16 we see that G is solenoidal, and since $t(\sum_p)$ is dense in $\sum_p$ for each $p \in \mathcal{P}$ we have $\overline{t(G)} = G$. It follows, however, from (b) that G cannot be p-thetic for any prime p. Another way to construct such a group is to observe that since J_p is torsion-free and $qJ_p = J_p$ for each prime $q \neq p$, $(J_p)_d$ contains a copy of $Q_p^{\#}$, the subgroup of Q consisting of rationals with denominator prime to p. It is then easy to check that if p and q are distinct primes, $G = (Q_p^{\#})\hat{} \times (Q_q^{\#})\hat{}$ is a compact solenoidal group having dense torsion subgroup, but G is p-thetic for no prime p.

(d) As a special case of 5.6 we see that for a given $p \in \mathcal{P}$ the group $\beta(Z(p^{\infty}))$ is a universal compact p-thetic group. Note that this group contains no elements of order q if q is a prime different from p. Cf. 5.35.

There are cases in which we may be sure that a compact connected group is p-thetic for some p. In fact, if the group is small enough and is not torsion-free such must be the case, as we see from the next result.

5.23 Proposition Let G be a compact and connected group whose dual has rank 1 (i.e., G has dimension 1; see §24.28 of [HR]). Then either $G \cong \hat{Q}$ or else G is p-thetic for some prime p.

Proof: If G is torsion-free it follows from P.24(a) that $G \cong \hat{Q}$. Otherwise, since G is divisible [P.28(b)], G must contain a subgroup H algebraically isomorphic to some $Z(p^{\infty})$-group [P.9(e)]. Thus $\bar{H}$ is compact p-thetic, and therefore (see, for example, 4.15) $(\bar{H})\hat{}$ is torsion-free. But it is clear that each quotient of the rank 1 group $\hat{G}$ by a nontrivial subgroup is a torsion group. Since $(\bar{H})\hat{} \cong \hat{G}/A(\hat{G}, \bar{H})$ we conclude that $A(\hat{G}, \bar{H})$ is trivial, that is, $G = \bar{H}$ and is therefore p-thetic. ∎

It would be very pleasant to have characterizations of the compact p-thetic groups similar to 5.5(e) for the compact monothetic groups and 5.6

[(a) ⇔ (c)] for the compact solenoidal groups. Indeed, 5.4 gives necessary and sufficient conditions for an abelian group to be a subgroup of T_d, and we have already seen in the proof of 5.16 that an abelian group is a subgroup of R_d iff it is torsion-free and has cardinality not exceeding the power of the continuum. But necessary and sufficient conditions for an abelian group to be a subgroup of some p-adic integer group appear to be unknown and probably quite difficult to formulate, so that intrinsic characterizations of compact p-thetic groups are not at hand.

We close the main body of this chapter by elaborating upon a comment made earlier. We have seen that if H is Z, R, or $Z(p^\infty)$ then an H-dense group G is either topologically isomorphic with H or else is compact. It turns out that there are no other groups $H \in \mathscr{L}$ (besides the compact ones, of course) for which this is so. This fact (whose proof we postpone until 10.2) may be regarded as a partial justification for concentrating on the H-dense groups for H one of these three types. Nevertheless, there are interesting questions about H-dense groups for other groups H, some of which the reader will find in the Miscellanea (see 5.38 and 5.39).

Miscellanea

5.24 Let $G \neq \{0\}$ in $\mathscr{L}$ be monothetic. If each nonzero element of G is a generator of G then G is cyclic of prime order.

5.25 (Armacost 1971c; compare with 5.10.) Let $G \in \mathscr{L}$ be monothetic. Then $n(G)$ is a subgroup of G iff G is one of the following: (a) $Z(p^n)$ for some $p \in \mathscr{P}$ and $n \in Z^+$, (b) J_p for some $p \in \mathscr{P}$, or (c) a compact connected group whose dual is of rank 1.

5.26 Let $G \in \mathscr{L}$ be monothetic. Call G *generator-homogeneous* iff for each x and y in $e(G)$ there exists a topological automorphism f of G such that $f(x) = y$. Then J_p is generator-homogeneous, but T and $\hat{Q}$ are not. For a related result see Corollary 3.3 of Comfort and Ross (1964).

5.27 Let $G \in \mathscr{L}$ be compact and totally disconnected. Then G is monothetic iff the subgroups of the form ker γ for $\gamma \in \hat{G}$ form a neighborhood base at 0.

5.28 A compact solenoidal group G may contain generators which lie on no one-parameter subgroup of G (cf. 5.19). [Take $G = \hat{Q}$. Construct a monomorphism $f \in \mathrm{Hom}(Q, T)$ such that f cannot be written in the form

$\exp(ig)$ for any $g \in \text{Hom}(Q, R)$ (§25.5 of [HR] may be helpful here). Identify f as a generator x of G. Then x lies on no one-parameter subgroup of G (cf. 4.31, or see §24.43 of [HR]).]

5.29 Let $G \in \mathscr{L}$ be compact and monothetic. Let S be the set of all $x \in G$ such that (a) $\gamma(x) \in t(T)$ for all $\gamma \in \hat{G}$ and (b) $\gamma(x) \neq 1$ for all nontrivial $\gamma \in t(\hat{G})$. Then $e(G) = e(c(G)) + S$. [The proof that $e(c(G)) + S \subseteq e(G)$ is straightforward. For the opposite inclusion, one may proceed as follows. Let $A \in \mathscr{L}_d$ be isomorphic to a subgroup of T_d. Let f be any monomorphism from A into T_d. Letting f' denote the restriction of f to $t(A)$, observe that $t(T_d)$ is divisible and use P.9(c) to extend f' to $\psi \in \text{Hom}(A, t(T_d))$. Then set $\phi = f - \psi$. Now for any $x \in e(G)$, set $A = \hat{G}$ and think of x as a monomorphism $f : A \to T_d$. Then construct ϕ and ψ as above. Identifying ϕ and ψ as elements c and s respectively in G, we have $x = c + s$. Finish by showing that $c \in e(c(G))$ and $s \in S$.] [Note: This result can be used to prove 5.12 without invoking 5.11. This was essentially the route followed by the author (Armacost 1971c) who was unblissfully unaware at the time of the work of Baayen and Helmberg (1965).]

5.30 [Armacost (to appear)] Theorem 5.11 remains true if T is replaced by any compact metrizable monothetic group K, but fails without the metrizability assumption.

5.31 (Khan 1976) A group $G \in \mathscr{A}$ is called *locally cyclic* iff each finitely generated subgroup of G is cyclic. It is well known that a nontrivial group $G \in \mathscr{A}$ is locally cyclic iff $r_0(G) + \max_{p \in \mathscr{P}} r_p(G)$ is 1 (Fuchs 1960, p. 34) and hence iff G is a subgroup of a quotient of Q [see Fuchs (1960, pp. 25–26)]. By way of generalization we call $G \in \mathscr{L}$ *locally monothetic* iff each closed compactly generated subgroup of G is monothetic. The groups Q, $\hat{Q}$, and the group D of 4.25(b) are all locally monothetic. Finally, $G \in \mathscr{L}$ is locally monothetic iff G is topologically isomorphic to a closed subgroup of a quotient by a closed subgroup of one of these three groups.

5.32 (Moskowitz 1967) Let $G \in \mathscr{L}$ and let H be a closed subgroup of G. If S is a one-parameter subgroup of G/H then there is a one-parameter subgroup S' of G such that $\pi(S') = S$, where $\pi : G \to G/H$ is the natural map. [Use P.21(b) and duality.]

5.33 Call an LCA group G *monogenic* iff there exists an element $x \in G$ having the property that whenever H is a closed subgroup of G such that

G/H is compact then $\pi(x) \in e(G/H)$, where $\pi: G \to G/H$ is the natural map. [This definition is due to L. Rubel in *Uniform distribution in locally compact groups*, Comm. Math. Helv. *39*, 253–258 (1965).] Clearly a compact group $G \in \mathcal{L}$ is monogenic iff G is monothetic, but there exist monogenic groups in $\mathcal{L}$ which are not monothetic, e.g., F_p. (In general, any densely divisible totally disconnected group $G \in \mathcal{L}$ is trivially monogenic.) The group R is not monogenic. For a complete description and classification of monogenic groups in $\mathcal{L}$ (which represents quite a tour de force) we refer the reader to Rajagopalan (1968a). [The monothetic and monogenic groups are of great interest in the theory of uniformly distributed sequences in locally compact groups. For an excellent survey of this topic the reader should consult Chapter 5 of Kuipers and Niederreiter (1974).]

5.34 (Armacost and Armacost 1972a) Let A be the discrete additive group of all rational numbers expressible in the form m/n with $m \in Z$ and n a square-free integer. Then $\hat{A}$ is p-thetic for all primes p, but is not topologically isomorphic to any torus [cf. 5.22(a)].

5.35 (Armacost and Armacost 1972a) Let $p \in \mathcal{P}$ be fixed and let G be a compact p-thetic group. Then $t(G)$ is a p-group iff $\hat{G}$ is isomorphic to a pure subgroup of $(J_p)_d$. (For other related results on p-thetic groups, see the paper referred to.)

5.36 (Armacost and Bruner 1973) If $G \in \mathcal{L}$ is not totally disconnected G is an S-group (see 1.25) iff G is a compact group of dimension 1 which is p-thetic for each prime p (cf. 5.23).

5.37 The product of two compact monothetic groups need not be monothetic. However, the product of two compact solenoidal groups must be solenoidal. What about the product of two compact p-thetic groups?

5.38 Let $p \in \mathcal{P}$ be fixed. A group $G \in \mathcal{L}$ is F_p-dense iff either $G \cong F_p$ or $G \cong Z(p^{\infty})$ or G is a compact solenoidal group. [If $f \in \mathrm{Hom}(F_p, G)$ has dense image then $f^* \in \mathrm{Hom}(\hat{G}, \hat{F}_p)$ is one-one, whence $\hat{G}$ must be totally disconnected. Letting K be a compact open subgroup of $\hat{G}$, conclude that if $K \neq \{0\}$ then f^* is a topological isomorphism from $\hat{G}$ onto a nontrivial closed subgroup of $\hat{F}_p$ ($\cong F_p$), whence $\hat{G} \cong J_p$ or $\hat{G} \cong F_p$. If $K = \{0\}$ then $\hat{G}$ is discrete, and since F_p and R are algebraically isomorphic (both being torsion-free divisible groups of cardinality c) we may apply 5.16. For the converse we note that a compact solenoidal group is F_p-dense for any $p \in \mathscr{P}$, since an injection from $\hat{G}$ into R_d gives rise to a continuous monomorphism from $\hat{G}$ into F_p.]

5.39 (Armacost and Armacost 1972b) This example deals with Q-dense groups.

(a) The groups Q, $Z(p^\infty)$, and R are obviously Q-dense.

(b) Any connected monothetic group is Q-dense. In particular, $(\hat{Q})^c$ is Q-dense.

(c) The groups F_p are Q-dense, and more generally, so is the group D of 4.25(b).

(d) The group $H = R \times D \times \hat{Q}^c$ is Q-dense, but is the "largest" Q-dense group in this sense: A group $G \in \mathscr{L}$ is Q-dense iff either $G \cong Q$ or else G is a quotient of H by a closed subgroup. Note in particular that since H is divisible, so is every Q-dense group.

(e) It may be proved from the last statement in (d) that a group $G \in \mathscr{L}$ containing a dense divisible subgroup of finite rank is necessarily divisible. The conclusion fails, however, if "finite" is replaced by "countable."

5.40 Call a group $G \in \mathscr{L}$ *polythetic* iff G contains a dense finitely generated subgroup. [The name seems to be due to G. H. Meisters. See Notices of the American Mathematical Society *19*, A–693 (1972), although such groups had been considered earlier, e.g., by G. Helmberg in Pac. J. Math. *8*, 227–241 (1958).] Polythetic groups have been studied by Takamatsu (1976). The following results are based on Muhin (1976).

(a) A compact connected polythetic group G is necessarily monothetic. [G is polythetic iff for some $n \in Z^+$ there exists $f \in \mathrm{Hom}(Z^n, G)$ having dense image. Hence $f^* \in \mathrm{Hom}(\hat{G}, T^n)$ is one-one. In particular $w(G) = w(\hat{G}) = |\hat{G}| \leq \mathfrak{c}$, so we may apply 5.5(c).]

(b) A compact torsion-free polythetic group is a product of $\mathfrak{m}$ copies of $\hat{Q}$ (where $0 \leq \mathfrak{m} \leq \mathfrak{c}$) and finitely many copies of J_p for various primes p.

(c) If G is polythetic and $b(G) = G$ then G is compact.

(d) Let G be polythetic and have compact identity component. Then $G \cong K \times Z^n$, where K is compact (and of course polythetic) and $n \in Z^{+0}$. [By P.17(c), $b(G)$ is open in G. Hence $G/b(G)$ is a discrete torsion-free polythetic group, i.e., has the form Z^n. By P.10(b), $b(G)$ is algebraically a direct summand of G, and since $b(G)$ is open, it is also a topological direct summand (cf. 6.8 below). Hence $G \cong b(G) \times Z^n$. Since $b(G)$ is a quotient of G, it must be polythetic too, so we finish by invoking (c).]

(e) The group R^n is polythetic for any $n \in Z^+$.

(f) A group $G \in \mathscr{L}$ is polythetic iff $G \cong R^n \times Z^m \times K$, where m and n are in Z^{+0} and K is compact polythetic. [Apply P.29 and (d).] (In particular, a polythetic group is compactly generated.)

5.41 Peterson (1973) shows that if $G \in \mathscr{L}$ is compact and monothetic, then every subgroup H of G having finite index is open in G. Consequently, if G is also totally disconnected, a character γ on G is continuous iff γ has finite order. Note that the same conclusions hold if "monothetic" is replaced by "polythetic." [Observe that $A(\hat{G}, nG)$ is finite for each $n \in Z^{+0}$.]

5.42 The study of H-dense groups leads naturally to considerations of the following sort: If $G \in \mathscr{L}$, now small (in the sense of cardinality) can dense subgroups or dense subsets of G be? We refer the interested reader to §24.31 of [HR] and to the following papers: Kakutani (1943), Hartman and Hulanicki (1958), Itzkowitz (1972), and Comfort and Itzkowitz (1977).

6

Splitting Problems

". . . to make all split." [Shakespeare, *A Midsummer Night's Dream*, Act 1, Scene 2]

To analyze the structure of a given LCA group it is generally very pleasant to be able to decompose the group into a product of smaller groups. Sometimes a decomposition in the algebraic sense comes right away to mind, but such decompositions are not usually satisfactory unless they are also topological.

6.1 *Definition* Let $G \in \mathcal{L}$ and let H be a closed subgroup of G. We say that H *splits* from G iff H is a topological direct summand of G in the sense of P.16.

We have not hitherto made much use of topological direct summands, although there have been occasions when some proofs could have been abbreviated by using the results of this chapter. We make up for this now by turning splitting problems into a minor obsession.

6.2 *Remarks* (a) If H and K are closed subgroups of $G \in \mathcal{L}$ such that $G = H \dotplus K$, then the isomorphism ϕ of P.16 from $H \times K$ onto G defined by $\phi(h,k) = h + k$ is automatically continuous, as may be proved directly. Hence, to prove that $G = H \oplus K$ we need only verify that ϕ is open. In particular, if H and K are both σ-compact, then so is $H \times K$, whence by P.30(b), ϕ is automatically open. Actually, and perhaps somewhat surprisingly, it suffices to know merely that H (or K) is σ-

compact (see 6.5), but ϕ may fail to be open if neither H nor K is σ-compact (see 6.3).

(b) If $G = H \oplus K$ then of course K splits from G as well. We call K a direct summand *complementary* to H. Although K is not in general uniquely determined, it is determined up to topological isomorphism, since we clearly have $K \cong G/H$.

6.3 Example Set $G = R \times R_d$, and let H be the diagonal subgroup $\{(r,r):r \in R\}$. It is clear that H is a closed (even discrete) subgroup of G. Let $K = \{(0,r):r \in R\}$. Clearly K is also a closed (discrete) subgroup of G. It is straightforward to verify that $G = H \dotplus K$. But $H \times K$ is discrete, so the map ϕ of P.16 cannot be open, i.e., $G \neq H \oplus K$. [We cannot conclude from this, however, that H does not split from G. We chose K to show that ϕ may not be open if neither H nor K is σ-compact. But if we set $K' = \{(r,0):r \in R\}$, it is an easy matter to verify that $G = H \oplus K'$, so that H does split from G.]

Next, consider the following situation. Let $G \in \mathscr{L}$ have a closed subgroup H. Suppose further that $G \cong H \times G_0$ for some $G_0 \in \mathscr{L}$. It is very tempting to conclude in an unthinking moment that H splits from G. We cannot, of course, conclude this even in the realm of discrete abelian groups, but we think the following example [from Ahern and Jewett (1965)] more illuminating. Besides, we shall want to cite it later (see before 6.21).

6.4 Example Think of T as $[0,1)$ with addition modulo Z and of $Z(p^\infty)$ as the subgroup (taken discrete) of elements of the form m/p^n. Set $G = Z(p^\infty) \times T$ and $H = \{(px,x):x \in Z(p^\infty) \subseteq T\}$. It is clear that H [being a nontrivial quotient of $Z(p^\infty)$] is algebraically isomorphic to $Z(p^\infty)$ and that H is a discrete (hence closed) subgroup of G. Thus we have $H \cong Z(p^\infty)$ and $G \cong H \times T$. Nevertheless, H does not split from G. For suppose so and that K is a complementary direct summand. Then we have $H \cong G/K$, and since H is discrete, K must be open in G. Hence [P.27(c)] we have $c(G) \subseteq K$, so that $c(G) \cap H = \{0\}$. But if we pick any $x \neq 0$ in $Z(p^\infty)$ such that $px = 0$, then the nonzero element $(0,x) = (px,x)$ belongs both to $c(G) = \{0\} \times T$ and to H, a contradiction. Therefore H does not split from G. [In fact, since G is σ-compact, each closed subgroup of G is σ-compact as well, whence it follows from 6.2(a) that for no closed subgroup K of G do we have $G = H \dotplus K$.] Compare this example with 6.32.

We now prove an assertion made in 6.2(a). Although we shall have no need for the following proposition in the sequel, we include it here because of its intrinsic interest. The result is from Fulp and Griffith (1971), but we give a different proof.

6.5 *Proposition* Let H be a closed σ-compact subgroup of $G \in \mathscr{L}$. If G has a closed subgroup K such that $G = H + K$, then $G = H \oplus K$.

Proof: By 6.2(a) we need only show that the map ϕ of P.16 from $H \times K$ onto G is open, or equivalently that ϕ^{-1} is continuous. It suffices to show that if $\{x_i\}_{i \in I}$ is a net converging to 0 in G (where I is a directed set), then the net $\{\phi^{-1}(x_i)\}_{i \in I}$ converges to the zero of $H \times K$. Set $\phi^{-1}(x_i) = (h_i, k_i)$, so that $h_i + k_i = x_i$ for each $i \in I$. We will be done if we show that $h_i \to 0$ in H, for then $k_i = x_i - h_i$ will converge to 0 in K, so that $(h_i, k_i) \to (0, 0)$ in $H \times K$. To show then that $h_i \to 0$ in H we define $\psi : H \to G/K$ by the rule $\psi(h) = h + K$ for each $h \in H$. It is straightforward to verify that ψ is a continuous monomorphism from H onto G/K. But since H is σ-compact, P.30(b) shows that ψ is open, i.e., ψ^{-1} is continuous. Let π_1 be the projection from $H \times K$ onto H and let π_2 be the quotient map from G onto G/K. Then the following diagram (the first, but not quite the last, in the book) commutes:

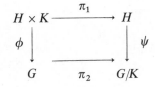

Now since $x_i \to 0$ we have $\pi_2(x_i) \to 0$ in G/K. But $\pi_2(x_i) = (\pi_2 \circ \phi)(h_i, k_i) = (\psi \circ \pi_1)(h_i, k_i) = \psi(h_i)$. But since ψ^{-1} is continuous, we have $h_i \to 0$ in H, as desired. ∎

A very simple and immensely useful criterion for splitting [due to Weil (1941, 1951) and also proved in §6.22(a) of [HR]] is contained in the next result.

6.6 *Theorem* Let H be a closed subgroup of $G \in \mathscr{L}$. Then H splits from G iff there exists $f \in \mathrm{Hom}(G, H)$ such that $f(h) = h$ for all $h \in H$. In this case $K = \ker f$ is a complementary direct summand.

Proof: Assume that such an f exists. Then clearly $H \cap K = \{0\}$. Moreover, each $x \in G$ can be written as $f(x) + [x - f(x)] \in H + K$, so we have $G = H + K$. Thus we need only show the continuity of ϕ^{-1}, where $\phi : H \times K \to G$ is as in P.16. Let $\{x_i\}_{i \in I}$ be a net in G converging to 0. Then $f(x_i) \to 0$ in H and $(x_i - f(x_i)) \to 0$ in K, so $\phi^{-1}(x_i) = (f(x_i), x_i - f(x_i)) \to (0, 0)$ in $H \times K$. Hence ϕ^{-1} is continuous at 0 in G and therefore everywhere continuous. Thus we have $G = H \oplus K$. For the converse, if ϕ is a topological isomorphism, let π be the projection from $H \times K$ onto H. Then $f = \pi \circ \phi^{-1}$ belongs to $\mathrm{Hom}(G,H)$ and satisfies $f(h) = h$ for each $h \in H$. ∎

6.7 Remarks (a) It is evident that if H is a subgroup of $G \in \mathscr{L}$ and if there exists $f \in \mathrm{Hom}(G,H)$ such that $f(h) = h$ for each $h \in G$, then H is automatically closed. Furthermore, it may be proved directly that f is an open mapping from G onto H.

(b) Suppose that H is a closed subgroup of $G \in \mathscr{L}$ and that $g \in \mathrm{Hom}(G,H)$. If the restriction of g to H is a topological automorphism of H, then H splits from G. For if we let g_0 be the restriction of g to H then the mapping $f = g_0^{-1} \circ g$ satisfies $f \in \mathrm{Hom}(G,H)$ and $f(h) = h$ for all $h \in H$.

We now draw a number of simple and useful conclusions from Theorem 6.6. The first of these is certainly well known and almost too simple to mention.

6.8 Corollary Let H and K be closed subgroups of $G \in \mathscr{L}$ and suppose that either H or K is open. Then if $G = H + K$ we have $G = H \oplus K$.

Proof: Assume for definiteness that K is open in G. Let $f : G \to H$ be defined by $f(x) = h$, where x is written uniquely as $h + k$ with $h \in H$ and $k \in K$. Then f is a homomorphism satisfying $f(h) = h$ for each $h \in H$, and since $\ker f = K$ is open, we see that f is continuous. Therefore by 6.6 we have $G = H \oplus K$. ∎

The next corollary is a slight generalization of §6.22(b) of [HR] [which in turn is from Weil (1941, 1951)]. We shall use the extra generality later, but the last sentence of the corollary is usually all we shall need.

6.9 Corollary Let H be a closed divisible subgroup of $G \in \mathscr{L}$ and let U be an open subgroup of G such that $H \subseteq U$. If H splits from U then H splits from G as well. In particular, an open divisible subgroup H of G splits from G.

Proof: Since H splits from U we see from 6.6 that there exists $f \in \text{Hom}(U,H)$ such that $f(h) = h$ for each $h \in H$. Since H is divisible we may [P.9(c)] extend f to a homomorphism $\bar{f}: G \to H$. But since $\bar{f}$ is continuous on the open subgroup U we see that $\bar{f} \in \text{Hom}(G,H)$, and since $\bar{f}$ is the identity on H it follows from (the other direction of) 6.6 that H splits from G. ∎

6.10 Corollary Let H be a closed subgroup of $G \in \mathscr{L}$. Then $A(\hat{G},H)$ splits from $\hat{G}$ iff H splits from G.

Proof: Assume that H splits from G. Then by 6.6 there exists $f \in \text{Hom}(G,H)$ such that f is the identity on H. Define $g \in \text{Hom}(G,G)$ by $g(x) = x - f(x)$ for each $x \in G$ and let $g^* \in \text{Hom}(\hat{G},\hat{G})$ be the adjoint of g. A straightforward computation reveals that g^* carries $\hat{G}$ into $A(\hat{G},H)$ and that g^* is the identity on $A(\hat{G},H)$. It then follows from 6.6 that $A(\hat{G},H)$ splits from $\hat{G}$. The converse may be proved by duality. ∎

We may insert here the following result borrowed from §25.30(b) of [HR].

6.11 Corollary Let $G \in \mathscr{L}$ be compact. If H is a closed subgroup of G such that G/H is torsion-free, then H splits from G.

Proof: By 6.10 it suffices to show that $A(\hat{G},H)$ splits from the discrete group $\hat{G}$. Now $(G/H)\hat{\ }$ is divisible [P.22(f) or P.24(a)]. Therefore $A(\hat{G}, H) \cong (G/H)\hat{\ }$ is a divisible subgroup of $\hat{G}$, whence by P.9(d), $A(\hat{G},H)$ splits from $\hat{G}$. ∎

We note for the record that the conclusion of 6.11 may fail if G is not compact. Indeed, there are familiar examples of discrete abelian groups A such that $t(A)$ does not split from A. [Take $A = \Pi_{p \in \mathscr{P}} Z(p)$, but with the discrete topology. It is easy to check that $A/t(A)$ is divisible, but A is itself reduced, so $t(A)$ cannot split from A.] For a generalization of 6.11 see 6.28.

We can use 6.11 to prove the following interesting splitting result of Itzkowitz (1968).

6.12 Proposition If $G \in \mathscr{L}$ is compact and connected, then $\overline{t(G)}$ splits from G. Moreover $G \cong G' \times (\hat{Q})^{\mathfrak{m}}$, where $G' \in \mathscr{L}$ has dense torsion subgroup and $\mathfrak{m}$ is some cardinal.

Proof: This can be proved by using 6.10, but we find Itzkowitz's proof more amusing. Let us show that $G/\overline{t(G)}$ is torsion-free. Suppose $nx \in \overline{t(G)}$

for $x \in G$ and $n \in Z^+$. There is a net $\{t_i\}$ of elements of $t(G)$ converging to nx. But G is divisible [P.28(b)], so $t(G)$ is divisible too. Hence for each t_i we find $s_i \in t(G)$ such that $t_i = ns_i$. But $\overline{t(G)}$ is compact, so the net $\{s_i\}$ has a subnet $\{s_j\}$ converging to some $s \in \overline{t(G)}$. Then $ns = \lim \overline{ns_j} = \lim t_j = nx$, so $n(x - s) = 0$. Therefore $x - s \in t(G)$, whence $x \in \overline{t(G)}$, proving that $G/\overline{t(G)}$ is torsion-free. It follows from 6.11 that $\overline{t(G)}$ splits from G. The remainder of the statement is then a consequence of P.24(a). ∎

It is unhappily the case that the identity component of an LCA group G need not split from G, even if G is compact. For example, let A be the discrete group mentioned right after the proof of 6.11 and let $G = \hat{A}$. Since by P.22(g) we have $A(\hat{G}, c(G)) = t(A)$, it follows from 6.10 and the fact that $t(A)$ does not split from A that $c(G)$ does not split from G. However, if G is torsion-free, $c(G)$ necessarily splits from G [§25.30(c) of [HR]]. Instead of proving this directly, we prove an interesting generalization due to Khan (1973a) [see also Khan (1980)].

6.13 Proposition If $G \in \mathscr{L}$ has closed torsion subgroup, then $c(G)$ splits from G. In particular, $c(G)$ always splits from a torsion-free $G \in \mathscr{L}$.

Proof: Denote $c(G)$ by C. First assume that C is compact. Since $t(C) = C \cap t(G)$ is closed, it follows from 6.12 that $t(C)$ splits from C. But then $t(C)$ is connected, whereas a torsion group must be totally disconnected (cf. 3.5). Therefore $t(C) = \{0\}$. Now by P.29, G has a compact open subgroup U. Since $C \subseteq U$ [P.27(c)] we see that if we set $H = t(G) \cap U$ we have $C \cap H = t(C) = \{0\}$. Now set $K = C + H \subseteq U$. Since $K = C \dotplus H$ and since C and H are both compact, we see [6.2(a)] that C splits from K. But U/K is torsion-free. [To see this, suppose $nu \in K$ for $u \in U$ and $n \in Z^+$. Write $nu = c + h$ for some $c \in C$ and $h \in H$. Since C is divisible (P.28(b)) we can write $c = nc'$ for some $c' \in C$, whence $n(u - c') \in H$. It follows that $u - c' \in H$ as well, so $u \in K$.] We conclude from 6.11 that K splits from U. But since C splits from K, an easy application of 6.6 shows that C splits from U. Finally, since C is divisible and U is open in G, we conclude from 6.9 that C splits from G.

The tricky part now being over, we write out the details for a general G. From P.29 we have $G = V \oplus G_0$, where $V \cong R^n$ for some $n \in Z^{+0}$ and G_0 has compact open subgroup and hence compact identity component. By 6.6 there are maps $g_1 \in \text{Hom}(G, V)$ and $g_2 \in \text{Hom}(G, G_0)$ such that g_1 (resp. g_2) is the identity on V (resp. G_0). Moreover, $t(G_0)$ is clearly closed in G_0, so by the first paragraph and 6.6 there exists

$g_3 \in \mathrm{Hom}(G_0, c(G_0))$ such that g_3 is the identity on $c(G_0)$. Since we clearly have $c(G) = V + c(G_0)$, the map $f = g_1 + g_3 \circ g_2$ belongs to $\mathrm{Hom}(G, c(G))$ and is the identity on $c(G)$. That is (6.6), $c(G)$ splits from G. ∎

The following consequence of 6.13 [for the special case $t(G) = \{0\}$] is a rewording of §3.27 of Robertson (1967).

6.14 Corollary Let $G \in \mathscr{L}$. If G is densely divisible then $b(G)$ splits from G. If in addition $c(G)$ is compact then $G = b(G) \oplus D$, where D is a discrete divisible subgroup of G topologically isomorphic to Q^{m*} for some cardinal $\mathfrak{m}$. Hence a group $G \in \mathscr{L}$ is densely divisible iff $G \cong R^n \times Q^{m*} \times L$, where $n \in Z^{+0}$, $\mathfrak{m}$ is a cardinal, and L is a densely divisible group satisfying $b(L) = L$.

Proof: If G is densely divisible then $\hat{G}$ is torsion-free (this is the easy half of 4.15). Therefore by 6.13, $c(\hat{G})$ splits from $\hat{G}$, whence by P.22(g) and 6.10, $b(G)$ splits from G, so we can write $G = b(G) \oplus D$ for some closed subgroup D of G. Now by P.17(c), $b(G)$ is open in G, so $D \cong G/b(G)$ is discrete. Clearly then D is divisible and torsion-free, so by P.9(e), D has the form Q^{m*}. In light of P.29 the rest of the statement is now clear. ∎

Up to this point it has been presumed when we say "H splits from G" that H is actually a closed subgroup of G. To avoid awkwardness in the sequel it will be convenient to extend the definition a bit.

6.15 Definition Let $\mathscr{C}$ be a class of LCA groups and let $H \in \mathscr{L}$. We say that H *splits from every* $G \in \mathscr{C}$ (or, more briefly, H *is splitting in* $\mathscr{C}$) iff whenever G is in $\mathscr{C}$ and contains a closed subgroup topologically isomorphic to H, then that closed subgroup splits from G.

It turns out that several familiar groups are splitting in various important classes of LCA groups. For example, we will show (6.23) that F_p is splitting in the class of torsion-free groups. Some groups, such as R and T, are even splitting in $\mathscr{L}$. In fact, we can say a bit more, as the next result shows. [This is given in Ahern and Jewett (1965), Moskowitz (1967), and Robertson (1968); the special case where G is a torus occurs as §25.31 of [HR].]

6.16 Theorem The groups $R^n \times T^m$ (where $n \in Z^{+0}$ and $\mathfrak{m}$ is any cardinal) are splitting in $\mathscr{L}$.

Proof: Let $G \in \mathcal{L}$ contain a closed subgroup $H \cong R^n \times T^m$. To expedite matters, write $R^n \times T^m$ as $S = \Pi_{i \in I} S_i$ where each S_i is either R or T and $S_i = T$ for all but finitely many i in the index set I. Let ϕ be a topological isomorphism from S onto H. For each $h \in H$ let $\phi^{-1}(h)$ be the "sequence" $(g_i(h))$. It is clear that $g_i \in \mathrm{Hom}(H, S_i)$ for each $i \in I$. Hence each g_i either belongs to $\hat{H}$ or is a continuous real character of H. In either case (P.21) each g_i may be extended to $\bar{g}_i \in \mathrm{Hom}(G, S_i)$. Now define $\psi \in \mathrm{Hom}(G, S)$ by setting $\psi(x) = (\bar{g}_i(x))$ for each $x \in G$. Finally, if we set $f = \phi \circ \psi$ we have $f \in \mathrm{Hom}(G, H)$ and $f(h) = h$ for each $h \in H$. Thus by 6.6, H splits from G, and since $G \in \mathcal{L}$ was arbitrary, we conclude that $R^n \times T^m$ is splitting in $\mathcal{L}$. ∎

We shall see later on (9.12) that the groups $R^n \times T^m$ are the only LCA groups having this "universal" splitting property. These groups also turn out to be the "injective" groups in $\mathcal{L}$ (see 9.15 and 9.17).

We now turn our attention to Q. In light of the previous remark, Q does not split from every LCA group. The following example is from Robertson (1968).

6.17 Example The group Q is not splitting in $\mathcal{L}$. To see this, set $G = R \times (Q/Z)$. Letting π denote the quotient map from Q onto Q/Z, we set $H = \{(q, \pi(q)): q \in Q\} \subseteq G$. It is readily verified that H is a discrete (hence closed) subgroup of G and that $H \cong Q$. But H does not split from G. For if so there exists by 6.6 some $f \in \mathrm{Hom}(G, H)$ such that $f(h) = h$ for all $h \in H$. But since H is totally disconnected we have $f(R \times \{0\}) = \{(0,0)\}$. However, $(1, \pi(1))$ belongs both to $R \times \{0\}$ and to H so $(0,0) = f((1, \pi(1))) = (1, \pi(1))$, an absurdity. (Cf. 6.31.)

The "problem" with G in the foregoing example is that $c(G)$ is not compact, as the next result (Ahern and Jewett 1965, Robertson 1968) shows.

6.18 Proposition Let $\mathscr{C}$ be the class of all $G \in \mathcal{L}$ such that $c(G)$ is compact. Then for any cardinal $\mathfrak{m}$ the group $Q^{\mathfrak{m}*}$ is splitting in $\mathscr{C}$.

Proof: Let $H \cong Q^{\mathfrak{m}*}$ be a closed subgroup of $G \in \mathscr{C}$. By P.17(c), $b(G)$ is open in G, so $G/b(G)$ is discrete. Now let π be the quotient map from G onto $G/b(G)$. Since $H \cap b(G) = \{0\}$ we see that π is one-one on H. If ϕ denotes the restriction of π to H, then ϕ is automatically a topological isomorphism from H onto $\pi(H)$. But $\pi(H)$ is divisible, so by P.9(d) and 6.6

there exists $g \in \mathrm{Hom}(G/b(G), \pi(H))$ such that g is the identity on $\pi(H)$. Setting $f = \phi^{-1} \circ g \circ \pi$ we have $f \in \mathrm{Hom}(G, H)$ and $f(h) = h$ for each $h \in H$. Thus by 6.6, H splits from G. ∎

For a generalization of this proposition see 6.31. We now investigate the splitting properties of $\hat{Q}$.

6.19 Example $\hat{Q}$ is not splitting in $\mathscr{L}$. By P.5(d), Q is a quotient of some free abelian group F (in fact, we can use $F = Z^{\aleph_0*}$). Taking F discrete and letting B be a subgroup of G such that $F/B \cong Q$, we set $G = \hat{F}$ and $H = A(\hat{F}, B)$. Then $H \cong \hat{Q}$ but since B clearly does not split from F it follows from 6.10 that H does not split from Q.

No such example can occur if G is torsion-free, as we now prove. The following result can be quickly deduced from Corollary 9 of Fulp (1972), but we give a direct proof, since we lack the elaborate machinery of Fulp's paper.

6.20 Proposition Let $\mathscr{C}$ be the class of torsion-free LCA groups. Then for any cardinal $\mathfrak{m}$ the group $(\hat{Q})^{\mathfrak{m}}$ is splitting in $\mathscr{C}$.

Proof: Let $G \in \mathscr{C}$ and let $H \cong (\hat{Q})^{\mathfrak{m}}$ be a closed subgroup of G. By P.27(e) we may find closed subgroups V and K of $c(G)$ such that $V \cong R^n$ for some $n \in Z^{+0}$, K is compact, and $c(G) = V \oplus K$. Clearly we have $H \subseteq K$, and since H is divisible, it follows from the torsion-freeness of K that K/H is also torsion-free. But then H splits from K (6.11). Since K splits from $c(G)$, and since $c(G)$ splits from G (6.13), it follows by repeated application of 6.6 that H splits from G. ∎

We now examine the groups $Z(p^\infty)$. In 6.4 we gave an example of a group G and a closed subgroup H of G such that $H \cong Z(p^\infty)$ but H does not split from G. This phenomenon cannot occur for totally disconnected groups: Robertson (1968) proved that $(Z(p^\infty))^{\mathfrak{m}*}$ for any cardinal $\mathfrak{m}$ is splitting in the class of totally disconnected groups. Actually, this is a special case of a much earlier result of Preston (1956), which we formulate as follows.

6.21 Proposition Any discrete divisible group D is splitting in the class $\mathscr{C}$ of totally disconnected LCA groups.

Proof: Let $H \cong D$ be a closed subgroup of a group $G \in \mathscr{C}$, and let K be a compact open subgroup of G. Then $K \cap H$ is at once discrete and compact, hence finite. Now by P.29(b) we can find an open subgroup L of K small enough to avoid the finitely many nonzero elements of $K \cap H$, whence we have $L \cap H = \{0\}$. Therefore the quotient mapping π from G onto the discrete group G/L maps H isomorphically onto the divisible subgroup $\pi(L)$ of G/L. The proof may be completed in just the same way as the proof of 6.18. ∎

The last groups we shall examine for their splitting properties are the groups F_p. It was shown in Robertson (1968) that F_p does not split from every $G \in \mathscr{L}$. Our example is a bit different from Robertson's.

6.22 Example For any $p \in \mathscr{P}$ the group F_p is not splitting in $\mathscr{L}$. We construct $G \in \mathscr{L}$ and a closed subgroup H of G such that $H \cong F_p$ but H does not split from G. We begin with the compact group $K = \Pi_{n=1}^{\infty} Z(p^n)$. Now it is easily seen that if L_n is the closed subgroup of J_p described in P.18(d) then $J_p/L_n \cong Z(p^n)$ for $n = 1, 2, \ldots$. Therefore there exist mappings $g_n \in \mathrm{Hom}(J_p, Z(p^n))$ such that $\ker g_n = L_n$ for each positive integer n. Define $g \in \mathrm{Hom}(J_p, K)$ by the rule $g(x) = (g_n(x))$ for each $x \in J_p$. Since $\cap_{n=1}^{\infty} L_n = \{0\}$ we see that g is one-one. Now let G be the minimal divisible extension of K as in P.31. By P.9(c) we may extend g to a homomorphism $\bar{g}: F_p \to G$, and since J_p is open in F_p, $\bar{g}$ is certainly continuous. It is moreover easy to verify that $\bar{g}$ is one-one. Now since G is totally disconnected [P.31(b)] it follows from 4.21 that $\bar{g}$ is a topological isomorphism from F_p onto a closed subgroup H of G. We claim that H does not split from G. For if so then (6.6) there exists $f \in \mathrm{Hom}(G, H)$ such that $f(h) = h$ for each $h \in H$. Since H is torsion-free f must annihilate $t(G)$. But $t(K)$ is dense in K, so $f(K) = \{0\}$. In particular, f annihilates the nontrivial subgroup $g(J_p)$ of K. But $g(J_p) \subseteq H$, so f cannot be the identity on H. This contradiction assures us that H does not split from G.

Robertson (1968) proved the very interesting result that the phenomenon described above cannot occur for torsion-free groups. Our proof of this result is rather different from his.

6.23 Proposition Let $\mathscr{C}$ be the class of torsion-free LCA groups. For any prime p the group F_p is splitting in $\mathscr{C}$.

Proof: We first make the following observation. Let $L \in \mathscr{C}$ be compactly generated and suppose that $x \in L$ satisfies $x \notin c(L)$. Then if

$\lim_{n \to \infty} p^n x = 0$ there exists $\phi \in \text{Hom}(L, J_p)$ such that $\phi(x) \neq 0$. To see this, we may assume [P.24(a) and P.26(a)] that L has the form $R^n \times Z^m \times (\hat{Q})^m \times S$, where n and m are in Z^{+0}, m is a cardinal, and S is a product of groups J_q for various primes q. Since $p^n x \to 0$ it is clear that x lies in $(\hat{Q})^m \times S$ and since $x \notin c(L)$ as well, x must have a nonzero component in a factor of S of the form J_p. We may now use the projection onto this factor to obtain ϕ.

Now to the main argument. We assume that $G \in \mathscr{C}$ and $H \cong F_p$ is a closed subgroup of G. From P.27(e) we see that $H \subsetneqq c(G)$, so there exists some $x \in H$ such that $x \notin c(G)$. Let L be a compactly generated open subgroup of G containing x [P.26(b)]. Now since $x \in H$ we have $\lim_{n \to \infty} p^n x = 0$, and since $x \notin c(L)$ we use the first paragraph to find $\phi \in \text{Hom}(L, J_p)$ such that $\phi(x) \neq 0$. Since F_p is divisible and L is open in G we can extend ϕ to $\bar{\phi} \in \text{Hom}(G, F_p)$ [P.9(c)]. Now $\bar{\phi}$ does not annihilate $\overline{\text{gp}(x)}$, which by P.25 is a compact subgroup of H. Therefore $\bar{\phi}(H)$ contains a nontrivial compact subgroup K. From P.18(d) we see that K is open in F_p, so $\bar{\phi}(H)$ is open (hence closed) in F_p. But since [P.18(d) again] no proper closed subgroup of F_p is divisible, the divisibility of $\bar{\phi}(H)$ implies that $\bar{\phi}(H) = F_p$. It is also evident that $\bar{\phi}$ is one-one on H [since each quotient of F_p by a proper closed subgroup is a torsion group, viz., $Z(p^\infty)$], so it follows from P.30(b) or 4.21 that the restriction of $\bar{\phi}$ to H (which we denote by ψ) is a topological isomorphism from H onto F_p. Setting $f = \psi^{-1} \circ \bar{\phi}$ we have $f \in \text{Hom}(G, H)$ and $f(h) = h$ for each $h \in H$. It now follows from 6.6 that H splits from G. ∎

We now use some of our findings to discuss indecomposable groups.

6.24 Definition A group G in $\mathscr{L}$ is said to be *indecomposable* iff the only closed subgroups of G which split from G are $\{0\}$ and G itself.

One can prove easily that the groups $Z, Q, Z(p^\infty), T, \hat{Q}, R, J_p$, and F_p are indecomposable. It is also evident from 6.10 that $G \in \mathscr{L}$ is indecomposable iff $\hat{G}$ is also indecomposable. The problem of describing or characterizing the indecomposable groups has not been solved even in the discrete case. In this case "most" indecomposable groups are torsion-free (see P.11). It has recently been shown by Shelah (1974) (improving on an earlier result of Fuchs) that there exist indecomposable groups in $\mathscr{L}_d$ of arbitrary infinite cardinality. Just by taking duals we can obtain a bewildering plethora of compact indecomposable groups. We shall content ourselves at present with two propositions about indecomposable

LCA groups. The first of these, dealing with nonreduced groups, appears to be new. The second is from Armacost (1976).

6.25 Proposition Let $G \in \mathscr{L}$ be indecomposable and not reduced. We have:

i) If G is a torsion group then $G \cong Z(p^{\infty})$ for some $p \in \mathscr{P}$.

ii) If G is torsion-free then G is topologically isomorphic to one of the following groups: R, Q, $\hat{Q}$, or F_p for some $p \in \mathscr{P}$.

Proof: (i) It follows from P.9(e) that G contains a subgroup D algebraically isomorphic to $Z(p^{\infty})$ for some $p \in \mathscr{P}$. But then $\bar{D}$ is p-thetic, and since (3.5) G is totally disconnected, it follows from 5.20 and 5.21 that $\bar{D} \cong Z(p^{\infty})$. Since G is indecomposable we see from 6.21 that $G \cong Z(p^{\infty})$.

(ii) From 4.23 we know that G contains a copy of one of the following groups: (a) R, (b) a group K which is a nontrivial quotient of $\hat{Q}$, (c) Q, (d) $Z(p^{\infty})$, or (e) F_p for some $p \in \mathscr{P}$. If (a) obtains then by 6.16 we have $G \cong R$. Assuming then that G contains no copy of R, we know that $c(G)$ is compact [cf. P.27(c) and P.29]. If (b) obtains then K must be $\hat{Q}$ itself, since [P.22(f)] $\hat{K}$ is a divisible subgroup of Q. But then 6.20 implies that K splits from G, whence $G \cong \hat{Q}$. If we have case (c) then $G \cong Q$ by 6.18. If we have neither (b) nor (c), then (e) must obtain, (d) being ruled out by the torsion-freeness of G. But then 6.23 tells us that $G \cong F_p$. ∎

6.26 Proposition Let $G \in \mathscr{L}$ be divisible and indecomposable. Then if G is not compact, G is topologically isomorphic to one of the following groups: R, Q, $Z(p^{\infty})$ or F_p for some $p \in \mathscr{P}$.

Proof: We first take a paragraph to prove a simple fact about minimal divisible extensions. If H_1^* and H_2^* are minimal divisible extensions (in the sense of P.31) of the LCA groups H_1 and H_2 respectively, then the divisible group $H_1^* \times H_2^*$ is a minimal divisible extension of $H_1 \times H_2$. Indeed, it is evident that $H_1 \times H_2$ contains the socle of $H_1^* \times H_2^*$ and that the quotient of $H_1^* \times H_2^*$ by $H_1 \times H_2$ is a torsion group. Since $H_1 \times H_2$ is open in $H_1^* \times H_2^*$ our assertion follows from P.31(c). In particular, if $H \in \mathscr{L}$ has indecomposable minimal divisible extension, then H must be indecomposable too.

Now to the proposition. By P.29 we see that either $G \cong R$ or else G has compact open subgroup H. Assuming the latter, we see that if $H = \{0\}$ then G is discrete, so P.9(e) implies that $G \cong Q$ or $G \cong Z(p^{\infty})$. Assuming henceforth that $H \neq \{0\}$, we let H^* be the minimal divisible extension of H. We claim that $G \cong H^*$. To see this, let $\iota : H \to G$ be the natural injection. Since G is divisible and H is open in H^* we may extend ι to $\bar{\iota} \in \mathrm{Hom}(H^*, G)$. Since $\bar{\iota}(H^*)$ is clearly an open divisible subgroup of G, 6.9 and the indecomposability of G show that $G = \bar{\iota}(H^*)$. Moreover, P.30(a) shows that $\bar{\iota}$ is an open mapping from H^* onto G. Finally, $\bar{\iota}$ is one-one, since $(\ker \bar{\iota}) \cap H = \ker \iota = \{0\}$ and H is essential in H^* [see P.8 and P.9(g)]. Therefore $\bar{\iota}$ is a topological isomorphism, so we have $G \cong H^*$, as asserted. But then H^* is indecomposable, whence by the first paragraph the compact group H (and hence $\hat{H}$) is indecomposable. Now $\hat{H}$ cannot be torsion-free, since then H would be divisible [P.28(b)], whence we would have $H = H^*$, forcing G to be compact. Therefore by P.11 and duality we have either $H \cong Z(p^n)$ or $H \cong J_p$. But it is then clear that either $H^* \cong Z(p^\infty)$ or $H^* \cong F_p$. Since $G \cong H^*$ we are done. ∎

6.27 Remarks (a) There are many examples of compact divisible [hence connected, by P.28(b)] indecomposable groups. In fact, by P.11 all compact indecomposable groups $G \in \mathscr{L}$ other than $Z(p^n)$ and J_p are connected, and Shelah's result alluded to before 6.25 shows that there are vast quantities of these. A more or less concrete example of a compact indecomposable group of cardinality $2^{\mathfrak{c}}$ is provided by our recent acquaintance $\beta(Z(p^\infty))$ of 5.22(d). [The fact that $\beta(Z(p^\infty))$ is indecomposable follows from the well-known result that $(J_p)_d$ is indecomposable (Fuchs 1960, p. 150, or Fuchs 1973, p. 123); that the group has cardinality $2^{\mathfrak{c}}$ follows from P.24(d).]

(b) One might ask what happens to Proposition 6.26 if we replace "divisible" by "densely divisible." We can proceed as follows. By 4.15, $\hat{G}$ is torsion-free, so by 6.25(ii) either $\hat{G}$ is reduced or $\hat{G}$ is one of the four groups $R, \hat{Q}, Q$, or F_p. Therefore a noncompact densely divisible indecomposable group G is either R, Q, or F_p or has reduced dual [e.g., $G = Z(p^\infty)$]. This result is not entirely satisfactory, inasmuch as 6.26 has an obvious converse, while the decomposable group $Z(p^\infty) \times Z(p^\infty)$ has reduced dual. The author would find an elucidation of the problem of describing the indecomposable reduced LCA groups (especially the nondiscrete torsion-free ones) to be of great interest.

(c) We recall from P.11 that an indecomposable group $G \in \mathscr{A}$ cannot be "mixed," that is, G must be either torsion or torsion-free. It would be of great interest to find an appropriate analogue of this theorem for LCA groups. The result certainly does not extend to $\mathscr{L}$ without change, since the indecomposable group T is mixed. [The same example shows that an indecomposable group need not be either a topological torsion group or a topologically torsion-free group (see the paragraph after Corollary

3.7).] However, all the indecomposable LCA groups G that we have met so far satisfy either $b(G) = G$ or $b(G) = \{0\}$. It is tempting to conjecture that this is true generally of indecomposable groups. By duality the conjecture may be phrased: An indecomposable LCA group is either connected or totally disconnected. Special cases follow from results already given. For instance, 6.25(ii) (or more simply 6.13) shows that an indecomposable torsion-free group must either be connected or totally disconnected. Without pausing to mention other special cases, we merely remark that the conjecture, if true, is perhaps rather difficult to prove.

Further results on splitting will be given in Chapter 9.

Miscellanea

6.28 Suppose that each element of $G \in \mathcal{L}$ is compact. If H is a closed subgroup of G such that G/H is compact and torsion-free, then H splits from G. (This generalizes 6.11.) $[A(\hat{G},H)$ is a discrete divisible subgroup of $\hat{G}$. Now use 6.21 and 6.10.]

6.29 (Cf. 6.11.) There exists a torsion-free group $G \in \mathcal{L}$ and a closed subgroup H of G such that G/H is torsion-free but H does not split from G. [For an example with discrete groups, use $G = F$ and $H = B$ in Example 6.19. Here is a more interesting example: Let $S \in \mathcal{L}$ be densely divisible but not divisible (cf. §4.16(d)). Let S^* be the minimal divisible extension of S. Take $G = (S^*)\hat{}$ and set $H = A(G,S)$. (This example appears again in 7.2.)]

6.30 It follows immediately from 6.11 that if $G \in \mathcal{L}$ is compact and $t(G)$ is finite then $t(G)$ splits from G. Does the same conclusion hold for arbitrary $G \in \mathcal{L}$? [Note that by P.10(b), $t(G)$ at least splits from G algebraically. But the author does not know the answer.]

6.31 Let $G \in \mathcal{L}$ contain a closed subgroup $H \cong Q^{m*}$ for some cardinal $\mathfrak{m}$. Then H splits from G iff $H \cap L = \{0\}$, where $L = b(G) + c(G)$ [cf. P.17(c)]. [This may be proved as in 6.18 (with L in place of $b(G)$); indeed, 6.18 is a special case of this. (Compare with 6.17, in which $H \cap L \neq \{0\}$, whence H cannot split. See also 6.33. For a very simple example in which 6.31 indicates splitting (while 6.18 does not apply) let $G = R \times Q$ and set $H = \{(q, q) : q \in Q\}$. Then $H \cong Q$ and $H \cap L = \{0\}$, so H splits from G.)]

6.32 Let D be a discrete divisible torsion group and let $G \in \mathscr{L}$ contain a closed subgroup $H \cong D$. Then H splits from G iff $H \cap c(G) = \{0\}$. [That $H \cap c(G) = \{0\}$ is necessary for splitting is evident. For the converse, first assume that $c(G)$ is compact. Then a modification of the proof of 6.21 shows that H splits from G. (For this part we do not require that D be a torsion group.) For the general case, use P.29. (Note that in Example 6.4 we had $H \cong Z(p^\infty)$ but $H \cap c(G) \neq \{0\}$, whence H could not split from G.)]

6.33 The previous result fails in general if the requirement that D be a torsion group is dropped. In fact, there exists $G \in \mathscr{L}$ having a closed subgroup $H \cong Q$ such that $H \cap c(G) = \{0\}$, but H does not split from G. [Set $G = R \times F_p \times (Q/Z)$, where p is any prime. Let f be a monomorphism from Q into F_p, and let π be the natural map from Q onto Q/Z. Define $H = \{(q, f(q), \pi(q)) : q \in Q\} \subseteq G$. Then $H \cong Q$ and $H \cap c(G) = \{0\}$, but H does not split from G (cf. 6.31).]

6.34 Let H and K be closed subgroups of $G \in \mathscr{L}$. Suppose that H and K both split from G. Under what circumstances can we conclude that $S = H + K$ also splits from G?

(a) S need not split from G. [Let $G = R \times Q$, $H = \{(r,r) : r \in Q\}$, $K = \{(t\sqrt{2}, t) : t \in Q\}$. Then $H \cong K \cong Q$ and both H and K split from G (cf. 6.31). However, $S = H + K$ is not even closed in G (it is a proper dense subgroup), much less does it split from G.]

(b) If $\mathrm{Hom}(H,K) = \{0\}$ then S is closed in G and splits from G. We also have $S = H \oplus K$. [By 6.6 there exist $f_1 \in \mathrm{Hom}(G,H)$ and $f_2 \in \mathrm{Hom}(G,K)$ such that f_1 (resp. f_2) is the identity on H (resp. K). The condition $\mathrm{Hom}(H,K) = \{0\}$ entails that $f_2(h) = 0$ for all $h \in H$. Setting $f = f_1 + f_2 - (f_1 \circ f_2)$, conclude that $f \in \mathrm{Hom}(G,S)$ and f is the identity on S. Hence by 6.7(a), S is closed in G and by 6.6, S splits from G. For the last statement, let ψ be the restriction of f_2 to S. Then $\psi \in \mathrm{Hom}(S,K)$, ψ is the identity on K, and $\ker \psi = H$. By 6.6, $S = K \oplus H$, whence $S = H \oplus K$.]

6.35 Let A denote either of the groups Q or F_p, where $p \in \mathscr{P}$. Let $G = A \times G_0$, where G_0 is any LCA group. If $H \cong A$ is a closed subgroup of G such that $H \subsetneqq \{0\} \times G_0$, then H splits from G. [Let π be the projection from G onto A and find $h \in H$ such that $\pi(h) \neq 0$. Now the topological automorphisms of A act transitively on the nonzero elements of A. (They are all multiplications. For F_p see §26.18(d) of [HR].) Hence find a topological isomorphism ϕ from A onto H such that $\phi(\pi(h)) = h$. Set

$f = \phi \circ \pi \in \mathrm{Hom}(G,H)$ and use 6.6. (Note by 6.4 that the result fails for $A = Z(p^\infty)$.)]

6.36 Let G be a nontrivial torsion-free topological p-group. The following are equivalent:

(a) There exists $\gamma \neq 1$ in $\hat{G}$ such that ker γ is compact.

(b) $G \cong (F_p)^n \times (J_p)^m$ where n is 0 or 1 and m is a cardinal. [(The argument for this is inspired by the proof of Lemma 7 of Rajagopalan and Soundararajan (1969).) Assume (a). If $o(\gamma) < \infty$ then G is compact, so use 2.8. If $o(\gamma) = \infty$ conclude that $G/\ker \gamma \cong Z(p^\infty)$. Set $H = A(\hat{G}, \ker \gamma)$. Then $H \cong J_p$ and is open in $\hat{G}$. Use 4.15 and 4.16(a) along with the aforementioned properties of H to show that $\hat{G}$ is divisible. Hence show that $\hat{G}$ contains an open subgroup of the form F_p. Then (b) follows from 6.8 and duality. Finally, it is straightforward to show that (b) $\Rightarrow$ (a).]

6.37 (Armacost 1972 and Flor 1978) Let G be a nontrivial LCA group. The following are equivalent:

(a) $|\mathrm{im}\ \gamma| = \aleph_0$ for each $\gamma \neq 1$ in $\hat{G}$.

(b) $G \cong Q^{m^*} \times G_0$, where $m \leq \aleph_0$ and G_0 is a densely divisible topological torsion group. [In Armacost (1972) the phrase "densely divisible," which is needed for the implication (b) $\Rightarrow$ (a), was omitted. The deficiency was repaired in Flor (1978).] (Use 4.15, 6.14, and 3.5.)

6.38 [Stated in Armacost (1971b); also see the paragraph before Definition 4.2.] Let H denote any of the groups T, R, $\hat{Q}$, Q, $Z(p^\infty)$, or F_p. If $G \in \mathscr{L}$ is indecomposable and $\mathscr{S}(G) = \mathscr{S}(H)$ then $G \cong H$. [This may be proved for each H individually by combining appropriate results from Chapter 4 and this chapter. For example, suppose $\mathscr{S}(G) = \mathscr{S}(F_p)$. Then $G \in \mathscr{S}(F_p)$, so G is totally disconnected and torsion-free. Since $F_p \in \mathscr{S}(G)$ there exists $f \neq 0$ in $\mathrm{Hom}(F_p, G)$. Since f must be one-one, use 4.21 to conclude that f is a topological isomorphism. But then $f(F_p)$ splits from G by 6.23, so $G \cong F_p$.]

7

Pure Subgroups

"Whatsoever things are pure, . . . think on these things." [St. Paul, *Philippians* 4:8]

In the theory of abelian groups the notion of pure subgroup ("serving subgroup" in the more conservative terminology) is of great importance. Direct summands are pure, while pure subgroups of certain types always split from their containing groups [see (a) and (b) of P.10]. Such splitting results do not generally carry over automatically, of course, to LCA groups, being often stubbornly resisted by topological complications. Nevertheless, there is a small body of interesting results about pure subgroups (usually assumed to be closed as well) of LCA groups. The study of purity in LCA groups dates from the work of Braconnier (1948) and Vilenkin (1946). An important paper devoted to the subject is Hartman and Hulanicki (1955), in connection with the results of which we shall find this chapter a convenient place to insert a discussion of the compactly generated groups and their duals, the groups "without small subgroups."

By way of review we recall (P.10) that a subgroup H of $G \in \mathscr{A}$ is pure iff $H \cap nG = nH$ for each $n \in Z^+$; that is, whenever $nx \in H$ for some $x \in G$ there exists $h \in H$ such that $nx = nh$. One may easily show the following useful fact (Fuchs 1970, p. 114): A subgroup H of G is pure iff H is p-pure (i.e., $H \cap p^k G = p^k H$ for $k = 1, 2, 3, \ldots$) for each $p \in \mathscr{P}$. Having refreshed our acquaintance with purity to this slight extent, we inform the reader that the other properties of purity that we shall need in this chapter are listed in P.10.

We now consider some naturally defined pure subgroups of G.

7.1 Examples Let $G \in \mathscr{L}$. (a) It is evident that $t(G)$ is pure in G. [It turns out, however, that $\overline{t(G)}$ need not be pure in G (see 7.14(b)).] Hence the closure of a pure subgroup need not be pure.

(b) $c(G)$ is pure in G. Indeed, $c(G)$ is divisible [P.27(e)].

(c) $b(G)$ is pure in G. In fact, we even have "strong purity," that is, $G/b(G)$ is torsion-free.

(d) [From Robertson (1968)] The p-component G_p (see 2.1) is pure in G for any prime p. To see this, first observe that if q is a prime different from p we have $q^k J_p = J_p$ for $k = 1, 2, 3, \ldots$ [cf. P.22(f)]. Suppose then that $q^k x \in G_p$ for some $x \in G$ and $k \in Z^+$. Now $\mathrm{gp}(q^k x)$ has the form J_p or $Z(p^n)$ (2.11), so in either case there exists some $\bar{y} \in J_p$ and $f \in \mathrm{Hom}(J_p, G)$ such that $f(\bar{y}) = q^k x$. Since $q^k J_p = J_p$ there exists $\bar{z} \in J_p$ such that $\bar{y} = q^k \bar{z}$. Thus $q^k x = q^k f(\bar{z})$, and since $f(\bar{z}) \in G_p$ and x was arbitrary, we conclude that $q^k G \cap G_p = q^k G_p$, i.e., G_p is q-pure. On the other hand, it is evident that G_p is p-pure, so G_p is pure in G. [If G is a topological torsion group we can conclude further (from Theorem 3.13) that G_p splits from G.]

(e) Various H-radicals (see 4.35) of G are pure in G. In fact, the H-radicals (a) through (d) of 4.35 are all pure in G. Note that (b) and (c) are special cases of this.

It is tempting to suppose from (b) and (c) above that if H is a closed subgroup of G then H is pure in G iff $A(\hat{G}, H)$ is pure in $\hat{G}$. This can be concluded under certain rather easily described conditions (see 7.10 below), but is not true in general. An example of a pure closed subgroup having nonpure annihilator is given in Hartman and Hulanicki (1955). The following example [from Khan (1973c)] describes a general situation in which this phenomenon occurs.

7.2 Example There is a group $G \in \mathscr{L}$ and a closed subgroup H of G such that H is pure in G, but $A(\hat{G}, H)$ is not pure in $\hat{G}$. Let S be any densely divisible but not divisible LCA group and let S^* be the minimal divisible extension of S. Set $G = (S^*)\hat{\ }$ and $H = A(G, S)$. (This is the same construction as in 6.29.) Then $(G/H)\hat{\ }(\cong S)$ is densely divisible, whence (4.15) G/H is torsion-free. Therefore H is pure in G, but $A(\hat{G}, H)$ is not pure in $\hat{G}$, since $\hat{G}$ $(\cong S^*)$ is divisible but $A(\hat{G}, H)$ $(\cong S)$ is not.

In light of this example it seems incumbent on us to seek necessary and sufficient conditions on a closed subgroup H of G for $A(\hat{G}, H)$ to be pure in $\hat{G}$. The property described in the next definition was first studied by, and is fundamental in the work of, Hartman and Hulanicki (1955).

7.3 *Definition* We say that a closed subgroup H of $G \in \mathcal{L}$ satisfies PHH (for "property of Hartman and Hulanicki") iff whenever $\gamma \in \hat{H}$ has order n there exists $\gamma' \in \hat{G}$ such that γ' extends γ and also has order n.

Note that if $\hat{H}$ is torsion-free [equivalently, H is densely divisible (4.15)] then H trivially satisfies PHH.

We now prove the basic fact about PHH [Hartman and Hulanicki (1955); see also §24.46 of [HR]].

7.4 *Proposition* Let H be a closed subgroup of $G \in \mathcal{L}$. Then H satisfies PHH iff $A(\hat{G},H)$ is pure in $\hat{G}$.

Proof: Assume that H satisfies PHH and suppose that $\gamma \in \hat{G}$ satisfies $\gamma^n \in A(\hat{G},H)$ for some $n \in Z^+$. We must find $\bar{\gamma} \in A(\hat{G},H)$ such that $\gamma^n = \bar{\gamma}^n$. Let γ_0 be the restriction of γ to H. Then $\gamma_0^n = 1$, whence γ_0 has order m dividing n. Since H satisfies PHH we find $\gamma' \in \hat{G}$ such that $o(\gamma') = m$ and $\gamma' = \gamma_0$ on H. Now set $\bar{\gamma} = \gamma(\gamma')^{-1}$. It is clear that $\bar{\gamma} \in A(\hat{G},H)$, and since $\bar{\gamma}^m = \gamma^m$ we have $\bar{\gamma}^n = \gamma^n$. Hence $A(\hat{G},H)$ is pure in $\hat{G}$.

For the converse, assume that $A(\hat{G},H)$ is pure in $\hat{G}$ and let $\gamma \in \hat{H}$ have order n. We must find $\gamma' \in \hat{G}$ such that $o(\gamma') = n$ and $\gamma' = \gamma$ on H. Let $\bar{\gamma}$ be any element of $\hat{G}$ which extends γ [P.21(a)]. Then $\bar{\gamma}^n \in A(\hat{G},H)$, whence by the purity of $A(\hat{G},H)$ there exists $\gamma_0 \in A(\hat{G},H)$ such that $\bar{\gamma}^n = \gamma_0^n$. Now set $\gamma' = \bar{\gamma}\gamma_0^{-1}$. It is clear that $\gamma = \gamma'$ on H. Moreover $(\gamma')^n = 1$, so $o(\gamma') \leqq n$. But $o(\gamma') < n$ is clearly impossible, so $o(\gamma') = n$ and H satisfies PHH. ∎

Our aim now is to give conditions ensuring that a closed pure subgroup H of G will have pure annihilator in $\hat{G}$. The preceding proposition then directs us to seek conditions ensuring that a closed pure subgroup will satisfy PHH. The following result seems to be the most general that is easily describable. [This is a generalization of Lemme 1 of Hartman and Hulanicki (1955); we replace their "compact" by "compactly generated" in part (b), one effect of which is to provide what we regard as a simpler proof of their main result, which we give below in 7.10.]

7.5 *Proposition* Let H be a closed pure subgroup of $G \in \mathcal{L}$. Suppose that either of the conditions (a) or (b) below obtains:
(a) nG is open in G for each $n \in Z^+$.
(b) H is compactly generated and nG is closed in G for each $n \in Z^+$.
Then H satisfies PHH, so that $A(\hat{G},H)$ is pure in $\hat{G}$.

Proof: Let $\gamma \in \hat{H}$ have order n. We must find $\gamma' \in \hat{G}$ such that $o(\gamma') = n$ and $\gamma' = \gamma$ on H. We shall first prove the result assuming condition (a). The same method will then be shown to work assuming condition (b), but this requires the verification of a few important points.

Assuming (a) then, let π be the quotient map from G onto the discrete group G/nG. Define a character γ_0 on $\pi(H)$ by the rule $\gamma_0(\pi(h)) = \gamma(h)$ for each $h \in H$. [This γ_0 is well defined; for if $\pi(h_1) = \pi(h_2)$ we have $h_1 - h_2 \in H \cap nG = nH$, and γ annihilates nH.] Now extend γ_0 arbitrarily to some $\bar{\gamma}_0 \in (G/nG)\hat{\ }$ [P.21(a)]. Since each element of G/nG has order dividing n, we have $(\bar{\gamma}_0)^n = 1$. Finally, define $\gamma' \in \hat{G}$ by setting $\gamma' = \bar{\gamma}_0 \circ \pi$. A straightforward computation now shows that γ' extends γ and has order n.

For the remainder of the proof we assume condition (b). We proceed just as above, but there is a hitch. Before, we knew that G/nG was discrete, so that γ_0 was automatically continuous. In the present situation nG is closed, so at least G/nG does not go outside $\mathcal{L}$. But we shall need to know that $\pi(H)$ is a closed subgroup of G/nG. In fact, $\pi(H)$ turns out to be compact, as we take the next paragraph to show.

Since H is compactly generated, we see from P.26(a) that there is a topological isomorphism f from a group $R^k \times Z^m \times K$ (where k and m are in Z^{+0} and K is compact) onto H. Set $A = f(R^k \times \{0\} \times \{0\})$, $B = f(\{0\} \times Z^m \times \{0\})$ and $C = f(\{0\} \times \{0\} \times K)$. Let π continue to denote the natural map from G onto G/ng. Since A is divisible we have $A \subseteq nG$, so $\pi(A) = \{0\}$. Since direct summands are pure and f is an isomorphism, we see that B is pure in H. But H is pure in G, and it follows easily (purity is "transitive") that B is pure in G. In particular, $B \cap nG = nB$. Now $\pi(B)$ is algebraically isomorphic to $B/(B \cap nG)$, so $\pi(B) \cong B/nB$. But the form of B shows that B/nB and hence $\pi(B)$ is finite. Finally, C is certainly compact. But $H = A + B + C$, whence $\pi(H)$ is the sum of a finite and a compact subgroup of G/nG, so $\pi(H)$ is compact, as desired.

Constructing the character γ_0 on $\pi(H)$ just as before, we show that γ_0 must be continuous. Since H is σ-compact (being compactly generated) and $\pi(H) \in \mathcal{L}$ we see from P.30(b) that π is an open mapping from H onto $\pi(H)$. To show that γ_0 is continuous we let U be a neighborhood of 1 in T and show that $\gamma_0^{-1}(U)$ is open in $\pi(H)$. Since by definition we have $\gamma_0 \circ \pi = \gamma$, a short computation reveals that $\gamma_0^{-1}(U) = \pi(\gamma^{-1}(U))$. Now by hypothesis γ is continuous on H, whence $\gamma^{-1}(U)$ is open in H. By what we have shown about π we know that $\pi(\gamma^{-1}(U))$ is open in $\pi(H)$, i.e., $\gamma_0^{-1}(U)$ is open in $\pi(H)$, as desired.

We now have γ_0 defined as a continuous character on the closed subgroup $\pi(H)$ of G/nG. Just as before, we can use P.21(a) to extend γ_0 to $\bar{\gamma}_0 \in (G/nG)\hat{}$. The rest of the proof is just as in the second paragraph. ∎

7.6 Corollary If $G \in \mathcal{L}$ is discrete or compact, then a closed subgroup H of G is pure iff $A(\hat{G}, H)$ is pure in $\hat{G}$.

Proof: If G is discrete, then (a) of Proposition 7.5 holds, while if G is compact (b) holds. In either case the purity of H then implies the purity of $A(\hat{G}, H)$. The converse follows by duality. ∎

Even though we are going to generalize Corollary 7.6 (see Theorem 7.10 below) we have stated it here both for convenience and also because it is pleasant to have some return on the investment of Proposition 7.5 before we take a brief detour. The detour consists of some facts about compactly generated groups and their duals, which we insert here both because of their intrinsic interest and because of their relevance to the main subject at hand. Our first result about compactly generated groups seems to appear first in Moskowitz (1967). It is such a basic fact (though not at all trivial) that it seems surprising that it should not have appeared in the literature earlier. Our proof is different from that of Moskowitz. For still other proofs see Morris (1977).

7.7 Proposition A closed subgroup H of a compactly generated group $G \in \mathcal{L}$ is compactly generated.

Proof: By P.26(a) we may assume $G = R^n \times Z^m \times K$, where m and n are in Z^{+0} and K is compact. Writing K for $\{0\} \times \{0\} \times K$, let π be the projection from G onto $G/K \cong R^n \times Z^m$. Either by direct argument or simply by invoking §5.18 of [HR] we find that π carries closed subsets of G onto closed subsets of G/K. Let π_0 be the restriction of π to H and set $L = \pi_0(H)$. Then L is a closed and hence locally compact subgroup of G/K [cf. P.14(a)]. Now since G is compactly generated it is certainly σ-compact, whence the closed subgroup H is σ-compact as well. Thus by P.30(b), π_0 is an open mapping from H onto L, so $L \cong H/\ker \pi_0 = H/(H \cap K)$. It is easy to show that our goal of proving that H is compactly generated will be reached if we show that $H \cap K$ is compact and L is compactly generated [cf. §5.39(h) of [HR]]. But $H \cap K$ is obviously compact, and since L is (topologically isomorphic to) a closed subgroup of $R^n \times Z^m$ and hence of R^{n+m}, we see from P.18(b) that L has the form $R^a \times Z^b$ and is therefore compactly generated. ∎

What do the duals of the compactly generated groups look like? It is clear from P.26(a) that they have the form $R^n \times T^m \times D$, where m and n are in Z^{+0} and D is discrete. Now it is easily seen that such groups have "no small subgroups" in the sense of the following definition.

7.8 Definition A group $G \in \mathscr{L}$ is said to have *no small subgroups* (or to be an NSS group) iff there exists a neighborhood U of 0 which contains no subgroup of G other than $\{0\}$.

It turns out, as we now show, that there are no NSS groups other than the ones already mentioned. The result is from Moskowitz (1967), but our proof is from Morris (1977).

7.9 Proposition The following are equivalent for a group $G \in \mathscr{L}$.
  (a) G is an NSS group.
  (b) $G \cong R^n \times T^m \times D$, where m and n are in Z^{+0} and D is discrete.
  (c) $\hat{G}$ is compactly generated.

Proof: Assume (a). If G is compact we see from P.24(c) that $G \cong T^m \times F$, where $m \in Z^{+0}$ and F is finite. In particular, a compact NSS group contains an open subgroup of the form T^m. For arbitrary G write $G \cong R^n \times G_0$, where G_0 contains a compact open subgroup K (P.29). Clearly K is also an NSS group, and we now see from the above that G contains an open subgroup of the form $R^n \times T^m$. It now follows from 6.9 (or 6.16) that (a) $\Rightarrow$ (b). The implications (b) $\Rightarrow$ (c) and (c) $\Rightarrow$ (a) being evident from P.26(a), we are done. ∎

We can now present the main result of Hartman and Hulanicki (1955) and of this chapter. Our formulation is a slight rewording of theirs, and our proof is rather different, as indicated before 7.5.

7.10 Theorem Let $\mathscr{C}$ denote the class of LCA groups which are either compactly generated or NSS groups. If H is a closed subgroup of $G \in \mathscr{C}$ then H is pure in G iff $A(\hat{G}, H)$ is pure in $\hat{G}$.

Proof: Let H be pure in G. If G is compactly generated it is clear from P.26(a) that nG is closed in G for each $n \in Z^+$. Moreover, Proposition 7.7 ensures that H is compactly generated. Therefore condition (b) of Proposition 7.5 holds, and $A(\hat{G}, H)$ is pure in $\hat{G}$. If on the other hand G is an NSS group, it follows from Proposition 7.9 [(a) $\Rightarrow$ (b)] that condition (a) of Proposition 7.5 holds, so again $A(\hat{G},H)$ is pure in $\hat{G}$. The converse follows from 7.9 and duality. ∎

Although we present no applications of Theorem 7.10 that require anything more than its special case Corollary 7.6, the theorem is important both on account of its depth and because it is a point of embarkation for further investigations. For instance, we may ask how far the class $\mathscr{C}$ may be extended without altering the conclusion. Indeed, Hartman and Hulanicki (1955) ask whether Theorem 7.10 is still valid if we replace $\mathscr{C}$ by the class of all groups of the form $D \times G$, where $D \in \mathscr{L}_d$ and $G \in \mathscr{C}$. As far as the author is aware, this question has never been answered.

We close the chapter with a discussion of two extreme cases encountered in considering how many pure closed subgroups an LCA group can be expected to possess. We answer the following two questions: (1) Which LCA groups have no closed pure subgroups except the trivial ones? [As in the discrete case (Fuchs 1970, p. 119) we call such groups *pure-simple*.] (2) For which LCA groups are all proper closed subgroups pure? Our answers are from Armacost (1974).

7.11 Proposition A group $G \in \mathscr{L}$ is pure-simple iff G is topologically isomorphic to one of the following: (a) a discrete group of rank 1, (b) J_p for some $p \in \mathscr{P}$, (c) F_p for some $p \in \mathscr{P}$, (d) R, or (e) a quotient of $\hat{Q}$ by a closed subgroup.

Proof: First observe that a discrete abelian group G is pure-simple iff G has rank 1, i.e., is isomorphic to a subgroup of Q or some $Z(p^\infty)$. [This is Exercise 7 on p. 119 of Fuchs (1970); it follows easily from P.10(a) and P.11.] We make essential use of this fact in the rest of the proof.

Now let G be an arbitrary pure-simple group. By 7.1(b) either $c(G) = G$ or $c(G) = \{0\}$. Assume the former. Since direct summands are pure we see from P.27(e) that either we have case (d) or G is compact. If G is compact then 7.6, P.28(b), the first paragraph, and duality lead us to case (e). We now assume that $c(G) = \{0\}$. If G is not torsion-free, then by P.11, G has a pure subgroup (indeed, an algebraic direct summand) H algebraically of the form $Z(p^n)$ or $Z(p^\infty)$, where $p \in \mathscr{P}$ and $n \in Z^+$. In the former case H is closed, so $G \cong Z(p^n)$ and we have case (a). In the latter case, Proposition 5.20 says that either $\bar{H}$ is compact and connected or $\bar{H} \cong Z(p^\infty)$. Since $c(G) = \{0\}$ we have the latter, whence $H = \bar{H} \cong Z(p^\infty)$ and we again have case (a). We are left with the case that G is totally disconnected and torsion-free. By 7.1(c) either $b(G) = \{0\}$ or $b(G) = G$. If the former holds, G is discrete [cf. P.17(c)], whence we are led to case (a). If $b(G) = G$ then $\hat{G}$ too is totally disconnected [P.22(g)], so G is a topological torsion group (3.5). But then Theorem 3.13 implies that G is a topological p-group for some $p \in \mathscr{P}$. Thus by 2.13 and 4.15, $\hat{G}$ is a densely divisible topological p-group. If G is trivial we of course have case (a), but otherwise 4.22 says that $\hat{G}$ contains a copy of $Z(p^\infty)$ or F_p. But then G has a closed subgroup H such that either $G/H \cong J_p$ or $G/H \cong F_p$. Since J_p and F_p are torsion-free, H is pure in G, so $H = \{0\}$ and we have cases (b) and (c). The proof is at last complete, except to remark that the converse is straightforward to verify. ∎

Now let us turn to the second of our two questions. To save a lot of unnecessary verbiage, let us agree to call a group $G \in \mathscr{L}$ *pure-full* iff each closed subgroup of G is pure in G. If $G \in \mathscr{L}$ is an *elementary p-group* (that is, $pG = \{0\}$ for some $p \in \mathscr{P}$) it is evident that all subgroups of G (closed or not) are pure. In fact, it is not hard to show that a discrete group $G \in \mathscr{L}$ is pure-full iff G is a weak direct product of (discrete) elementary p-groups. [This is Exercise 12 on p. 16 of Kaplansky (1969).] If G is not discrete, however, G may be pure-full without being a torsion group. For example, let G be the compact group $\Pi_{p \in \mathscr{P}} Z(p)$. By what we have already said, the discrete group $\hat{G}$ is pure-full, so by 7.6, G is pure-full as well. Note that each p-component G_p of G is elementary. The general situation is as follows.

7.12 Proposition A group $G \in \mathscr{L}$ is pure-full iff G is a local direct product of elementary p-groups with respect to compact open subgroups.

Proof: Let G be pure-full. It is evident that each closed subgroup of G is also pure-full. Since R is not pure-full we see [P.27(e)] that $c(G)$ is compact. But since $c(G)$ is pure-full it follows from 7.6 that the discrete dual of $c(G)$ is pure-full too, and what we have already said of the discrete case shows that $c(G) = \{0\}$. Furthermore, since Z is not pure-full we see from P.25 that $b(G) = G$, so G is a topological torsion group [P.22(g) and 3.5]. We now show that each p-component G_p is an elementary p-group. Indeed, each subgroup G_p of G is closed (3.8) and is therefore pure-full. Now pick any $x \in G_p$ and set $H = \overline{gp(x)}$. Then H has the form J_p or $Z(p^n)$ (2.11) and since H is pure-full it is evident that $H = \{0\}$ or $H \cong Z(p)$. It then follows that G_p is an elementary p-group. We conclude from Theorem 3.13 that G is a local direct product of elementary p-groups. For the converse note that if H is a closed subgroup of a group G having the form mentioned, then H_p is pure in G_p for each prime p. One can then show by applying Theorem 3.13 to H that H is pure in G. ∎

7.13 *Remarks* (a) The elementary p-groups are described in 2.25.

(b) Proposition 7.12 is closely related to a result stated in Čarin (1967) to the effect that the local direct products of elementary p-groups are the only LCA groups G having the following property: For each closed subgroup H of G there is a closed subgroup K of G such that $G = H + K$.

Miscellanea

7.14 The closure of a pure subgroup H of $G \in \mathscr{L}$ need not be pure in G. Apparently the first example of this phenomenon occurs in Braconnier (1948) and is given also in §24.46(c) of [HR]. Two other examples, in which H is not only pure but even more special, are as follows.

(a) The closure of a divisible subgroup H of G need not be pure in G. [Take any nondivisible group $S \in \mathscr{L}$ containing a dense divisible subgroup H (cf. 4.16(d)), and let G be the minimal divisible extension of S.]

(b) The closure of the torsion subgroup of G need not be pure in G. [Let A be a reduced group in $\mathscr{L}_d$ such that $\cap_{n=1}^{\infty} nA \neq \{0\}$ [P.9(h)] and set $G = \hat{A}$.]

7.15 Let $G \in \mathscr{L}$ be divisible and torsion-free. A compact subgroup H of G splits from G iff H is pure in G. [Use P.24(a) and 6.20.]

7.16 Let $G \in \mathscr{L}$ be compact. If H is a closed pure subgroup of G such that G/H is a torsion group, then H splits from G. [Use P.24(b), 7.6, P.10(b), and 6.10.]

7.17 (Leptin 1956) Let $G \in \mathscr{L}$ be compact. A closed subgroup H of G is pure in G iff H is an algebraic direct summand of G. [This is also shown in Hartman and Hulanicki (1955). Readers familiar with algebraically compact groups will see that the result is a direct consequence of §38.1 of Fuchs (1960). (We note that H need not be a topological direct summand of G, as we see from the example in the paragraph preceding Proposition 6.13. If, however, H is open in G, then of course H splits from G, since any homomorphism $f : G \to H$ which is the identity on H is automatically continuous. Another condition on H ensuring that H splits has already appeared in 7.16.)]

7.18 Let $G \in \mathscr{L}$. (a) For any $\gamma \in \hat{G}$ set $K(\gamma) = \cup_{n=1}^{\infty} \ker \gamma^n$. Then $K(\gamma)$ is a pure subgroup of G. [The subgroup $K(\gamma)$ has already put in a brief appearance in the proof of Proposition 5.9.]

(b) If G is not totally disconnected, then G contains a proper dense pure subgroup. [By P.28(a) and P.21(a) there is a surjective $\gamma \in \hat{G}$. For any such γ we have $K(\gamma) \subsetneqq G$. In showing that $K(\gamma)$ is dense, it may be helpful to use 8.4 below.]

(c) Open problem: Which groups G have no proper pure dense subgroups? (Compare with 2.26.)

7.19 Let p be a fixed prime and let G be a topological p-group. A closed subgroup H of G is pure in G if H is p-*pure* (i.e., $p^k G \cap H = p^k H$ for $k = 1, 2, 3, \ldots$). (Compare with 2.16.)

7.20 Theorem 7.10 does not remain valid if we replace the class $\mathscr{C}$ by the class of topological p-groups. [Let S be the local direct product of $\aleph_0$ copies of $Z(p^\infty)$ with respect to the compact open subgroups $Z(p)$. It is not hard to see that S is densely divisible but not divisible. Let S^* be the minimal divisible extension of S and set $G = (S^*)\hat{\ }$. Then G is a topological p-group (2.13 and 2.15). If we set $H = A(G,S)$, then H is pure in G but $A(\hat{G},H)$ is not pure in $\hat{G}$ (cf. 7.2). ($A(\hat{G},H)$ does, however, turn out to be "topologically pure" in $\hat{G}$, for which see 7.21 following.)]

7.21 (Vilenkin 1946) Let p be a fixed prime. A closed subgroup H of a topological p-group G is said to be *topologically pure* in G iff $H \cap \overline{p^k G} = \overline{p^k H}$ for $k = 1, 2, 3, \ldots$ (cf. 7.19). If H is any closed pure subgroup of a topological p-group G, then $A(\hat{G},H)$ is topologically pure in (the topological p-group) $\hat{G}$. [Vilenkin's proof of this (see his Theorem 22) goes through without the second countability restriction that he regularly imposes. Note that the result, in conjunction with 7.20, implies that a topologically pure subgroup need not be pure. It is also true (cf. Example 6 at the end of Vilenkin (1946)) that purity does not imply topological purity.]

7.22 Let us agree to call an LCA group G *character-pure* iff the group of continuous characters of G is a pure subgroup of the group of all characters of G. Discrete groups are trivially character-pure. It is moreover evident that if $\hat{G}$ is divisible (e.g., $G = R, \hat{Q}, J_p, F_p$) then G is character-pure. The group T is also character-pure (Rajagopalan 1968b, Lemma 12). [Rajagopalan proves this directly, but it is also a consequence of an interesting result due to Venkataraman (1970). Venkataraman defines a concept of "topological purity" (stronger than purity and different from that of 7.21). A corollary of his Theorem 1 is the following: If $G \in \mathscr{L}$ is σ-compact and nG is closed in G for each $n \in Z^+$, then G is pure in its Bohr compactification $\beta(G)$ (P.32). By this it is meant that the image of G under the mapping Φ of P.19(d) (the codomain being changed from $(\hat{G})\hat{}$ to $\beta(G)$) is a pure subgroup of $\beta(G)$. If we take $G = Z$ we obtain Rajagopalan's result. It would be of interest to make a detailed investigation of the character-pure LCA groups.]

7.23 (Khan 1973b) Call a closed subgroup H of $G \in \mathscr{L}$ *t-pure* in G iff whenever K is a closed subgroup of G such that $K \supseteq H$ and K/H is compactly generated, then H splits from K. [This definition is motivated by Theorem 28.4 of Fuchs (1970).] We have:
   (a) If H is a topological direct summand of G then H is t-pure in G.
   (b) t-pure subgroups of G are pure in G.
   (c) t-pure divisible subgroups of G split from G.

7.24 Open Problem: If every pure subgroup of $G \in \mathscr{L}$ is closed, must G be discrete? (Compare with 1.26.)

7.25 (Khan 1980) A group $G \in \mathscr{L}$ is an NSS group iff the compact subgroups of G satisfy the minimum condition. (This is just the dual of 1.28.)

7.26 (Morris 1972) It is clear from P.26(a) that a compactly generated group $G \in \mathscr{L}$ may be built up from R by repeatedly taking subgroups, quotients, and direct products. It turns out that any LCA group so obtained from R is necessarily compactly generated. See Morris' paper for the precise result.

8

Connectedness Properties

"Nec tamen omnimodis conecti posse putandum est omnia." [Lucretius, *De Rerum Natura*, II, 700–701]

Although the word "connected" has probably appeared on nearly half of the foregoing pages, we have not yet had much occasion to investigate the finer points of the subject of connectedness in LCA groups. We begin this chapter (by way of complementing P.28) with a few easy results about the relation between the connectedness of a group and mapping properties of its characters. Then we pass to a consideration of local and arcwise connectedness. We shall give dual formulations for both these properties.

We begin by investigating the converse of P.28(a): If $G \in \mathscr{L}$ and each $\gamma \neq 1$ in $\hat{G}$ maps G onto T, must G be connected? It is quite easy to prove that this is so, but first we need a simple lemma. [The results in 8.1 through 8.6 are from a paper of the author (Armacost 1972).]

8.1 Lemma Let $A \in \mathscr{L}_d$ and let a be any nonzero element of A. There exists $\gamma \in \hat{A}$ such that $\gamma(a) \neq 1$ and $|\mathrm{im}\ \gamma| \leq \aleph_0$.

Proof: By forming any divisible extension of A [e.g., as in P.9(g)] and recalling P.9(e) we may suppose that A is a subgroup of a group A' which is a weak direct product of copies of Q and various $Z(p^\infty)$ groups. Since each of these groups is a subgroup of T we can form the required γ by composing the injection from A into A' with the projection from A' onto an appropriate summand. ∎

8.2 *Proposition* A group $G \in \mathscr{L}$ is connected iff each $\gamma \neq 1$ in $\hat{G}$ is surjective.

Proof: Assume that each $\gamma \neq 1$ in $\hat{G}$ is surjective, but that G is not connected. If $U \neq G$ is an open subgroup of G [P.27(c)] and $x + U \neq U$ in the discrete group G/U, there exists by 8.1 some $\gamma \in (G/U)\hat{}$ such that $\gamma(x + U) \neq 1$ and $|\mathrm{im}\ \gamma| \leqq \aleph_0$. If $\pi : G \to G/U$ is the natural map, then $\gamma' = \gamma \circ \pi$ is a nontrivial element of $\hat{G}$ which cannot be surjective. This violates our assumption, so G is connected. The converse is just P.28(a). ∎

8.3 *Remark* We have actually shown that G is connected iff $|\mathrm{im}\ \gamma| > \aleph_0$ for each $\gamma \neq 1$ in $\hat{G}$. (Compare with 8.28.)

Now it is clear from P.27(e) that a connected LCA group is necessarily σ-compact. It then follows from P.30(b) that if $G \in \mathscr{L}$ is connected then each $\gamma \neq 1$ in $\hat{G}$ is an open mapping. The converse is true and follows easily from 8.2. Hence $G \in \mathscr{L}$ is connected iff each $\gamma \neq 1$ is an open mapping. This is a special case of the next result, which identifies the open characters $\gamma \in \hat{G}$ as the characters which, as elements of $\hat{G}$, are not compact.

8.4 *Proposition* The following are equivalent for $G \in \mathscr{L}$ and $\gamma \in \hat{G}$:
  (a) γ is an open mapping.
  (b) $\gamma(c(G)) \neq \{1\}$.
  (c) $\gamma \notin b(\hat{G})$.

Proof: For this proof let C denote $c(G)$. Assume (a). Then evidently γ is surjective, so $G/\ker \gamma \cong T$. If (b) does not hold, then we have $C \subseteq \ker \gamma$, whence $G/\ker \gamma \cong (G/C)/(\ker \gamma/C)$ (cf. §5.35 of [HR]). But then P.27(a) and (d) imply the absurdity that T is totally disconnected, whence we conclude (a) $\Rightarrow$ (b). Now assume (b) and let γ_0 be the restriction of γ to C. We have shown above that γ_0 must be an open mapping from C onto T. If then U is any neighborhood of 0 in G, $U \cap C$ is a neighborhood of 0 in C, so $\gamma(U \cap C) = \gamma_0(U \cap C)$ is a neighborhood of 1 in T. It then follows from P.30(a) that γ is an open mapping, i.e., (b) $\Rightarrow$ (a). Finally, the equivalence of (b) and (c) follows from P.22(g). ∎

8.5 *Corollary* Let $G \in \mathscr{L}$. Then G is connected iff each $\gamma \neq 1$ in $\hat{G}$ is open, while G is totally disconnected iff no $\gamma \in \hat{G}$ is open. ∎

8.6 *Corollary* The group $G \in \mathscr{L}$ is not totally disconnected iff there are sufficiently many open $\gamma \in \hat{G}$ to separate the points of G.

Proof: If G has any open continuous characters at all, it follows from 8.5 that G is not totally disconnected. Conversely, if G is not totally disconnected, then $b(\hat{G}) \neq \{1\}$ [P.22(g)]. Now it is evident that each element of $\hat{G}$ can be written as a product of two noncompact elements of $\hat{G}$, whence by 8.4 each $\gamma \in \hat{G}$ can be written as a product of open characters in $\hat{G}$. It then follows from P.19(a) there are sufficiently many open $\gamma \in \hat{G}$ to separate the points of G. ∎

Proposition 8.4 also has the following odd-sounding consequence.

8.7 *Corollary* Let $G \in \mathscr{L}$ and $\gamma \in \hat{G}$. Then γ is an open mapping iff ker γ is not open in G.

Proof: If γ is open it is obvious that ker γ is not open. For the converse, assume that ker γ is not open. Then by P.22(e) and (j) the subgroup $\overline{\mathrm{gp}(\gamma)}$ of $\hat{G}$ is not compact, i.e., $\gamma \notin b(\hat{G})$. It then follows from 8.4 that γ is an open mapping. ∎

8.8 *Corollary* Let $G \in \mathscr{L}$. Then G is connected iff ker γ is nonopen for each $\gamma \neq 1$ in G, and G is totally disconnected iff ker γ is open in G for each $\gamma \in \hat{G}$.

Proof: Both parts follow immediately from 8.4 and 8.7. [Note that the second part also follows immediately from P.28(c), P.27(c), and P.19(a).] ∎

We now study local connectedness in LCA groups. We say that a topological space is *locally connected* iff there is a base for the topology of the space consisting of open connected subsets. It is obvious that a discrete space is locally connected. A topological group G is locally connected iff there is a base at 0 of open connected subsets of G. Examples of locally connected LCA groups are R and T, and more generally $R^n \times T^m$, where $n \in Z^{+0}$ and $\mathfrak{m}$ is any cardinal. If $G \in \mathscr{L}$ is locally connected, then $c(G)$ is evidently open in G; since $c(G)$ is divisible [P.27(e)] we see from 6.9 that $c(G)$ splits from G. This leads to the following simple structure theorem for locally connected LCA groups [from Braconnier (1948), slightly sharpened as in Theorem 33 of Morris (1977)].

8.9 *Proposition* A group $G \in \mathscr{L}$ is locally connected iff $G \cong R^n \times K \times D$, where $n \in Z^{+0}$, K is compact, connected, and locally connected, and $D \in \mathscr{L}_d$.

Proof: We have already shown that if G is locally connected, then $c(G)$ is open and splits from G. From P.27(e) we see that $G \cong R^n \times K \times D$, where $n \in Z^{+0}, K$ is compact and connected, and $D \in \mathcal{L}_d$, and since K is a quotient of G, K is locally connected too. The converse follows from the fact that a product of finitely many locally connected spaces is again locally connected. ∎

We see from Proposition 8.9 that in studying local connectedness of LCA groups we may restrict our attention to compact connected groups. Local connectedness of such groups is a relatively rare phenomenon. Our first step in this direction is a fundamental result due to Pontryagin [1966, p. 260 (B)].

8.10 Proposition Let D be a discrete torsion-free abelian group of finite rank. Then the compact group $\hat{D}$ is locally connected iff D is finitely generated.

Proof: If D is finitely generated then [P.6(b)] $D \cong Z^n$ for some $n \in Z^{+0}$, so $\hat{D}$ is the locally connected group T^n.

Conversely, assume that $\hat{D}$ is locally connected. Let $\{d_1, \ldots, d_n\}$ be a maximal independent set in D and let H be the subgroup of D generated by $d_1, \ldots, d_n$. It is evident that D/H is a torsion group.

We shall require a moderately small open interval I about 0 in R, say $I = (-1/3, 1/3)$. Let E be the product I^n of n copies of I. For each $t = (t_1, \ldots, t_n) \in E$ we define $\gamma_t \in \hat{D}$ in the following way: If $d \in D$ there exist integers $m_1, \ldots, m_n$ and another integer $m \neq 0$ such that $md = m_1 d_1 + \cdots + m_n d_n$. We then set $\gamma_t(d) = \exp[2\pi i m^{-1}(m_1 t_1 + \cdots + m_n t_n)]$. It is directly verified that γ_t is well defined and is indeed a character of D (cf. §24.27 of [HR]). Let L be the topological subspace of $\hat{D}$ consisting of all γ_t for $t \in E$. Bearing in mind the fact that the topology of $\hat{D}$ is that of pointwise convergence on D, one verifies directly that the map $f: E \to L$ defined by $f(t) = \gamma_t$ is a homeomorphism from E onto L.

Now let V be the neighborhood $\{\exp(2\pi i r): r \in I\}$ of 1 in T and set $W = \{\gamma \in \hat{D}: \gamma(d_i) \in V \text{ for } i = 1, \ldots, n\}$. Then W is open in $\hat{D}$ and is hence locally connected as a topological space with its topology inherited from $\hat{D}$. Consider $A(\hat{D}, H)$ also as a topological space with its relative topology. We are going to show that W is homeomorphic to the product space $L \times A(\hat{D}, H)$. To this end pick any $\gamma \in W$. Then for $i = 1, \ldots, n$ we have $\gamma(d_i) \in V$, so for each such i there exists a unique $t_i \in I$ such that $\gamma(d_i) = \exp(2\pi i t_i)$. Set $t = (t_1, \ldots, t_n) \in E$ and let γ_t be the character $f(t)$

defined in the preceding paragraph. Now γ and γ_t agree at each d_i, so they agree on H. Hence $\gamma\gamma_t^{-1} \in A(\hat{D},H)$, so we can write $\gamma = \gamma_t\eta$ for some $\eta \in A(\hat{D},H)$. It is readily verified that this representation of γ is unique, which means that we obtain a well-defined map ϕ from W into $L \times A(\hat{D},H)$ by setting $\phi(\gamma) = (\gamma_t, \eta)$ for each $\gamma \in W$. It is evident that ϕ is a one-one mapping from W onto $L \times A(\hat{D},H)$, and by using the fact that f is a homeomorphism from E onto L we can verify directly that ϕ is a homeomorphism. In particular, $L \times A(\hat{D},H)$ is locally connected.

At last the proof is poised on the brink of completion. Since $A(\hat{D},H)$ is the open continuous image of the locally connected space $L \times A(\hat{D},H)$, we know that $A(\hat{D},H)$ is locally connected too. But since $(D/H)\hat{} \cong A(\hat{D},H)$, we see that $(D/H)\hat{}$ is locally connected. However, $(D/H)\hat{}$ is also totally disconnected [P.28(b)]. It follows that $(D/H)\hat{}$ is discrete, and since it is compact, it must be finite. But then D/H is finite [P.20 (a)]. Since H is finitely generated it follows immediately that D is finitely generated, as we wished to show. ∎

8.11 Remarks (a) Proposition 8.10 implies immediately that our familiar friend $\hat{Q}$ is not locally connected.

(b) Proposition 8.10 may also be phrased (taking §24.28 of [HR] into account) as follows: A compact connected finite dimensional group $K \in \mathcal{L}$ is locally connected iff $K \cong T^n$ for some $n \in Z^{+0}$.

Proposition 8.10 will be useful in showing that the property described in the next definition is dual to that of local connectedness.

8.12 Definition We say that $G \in \mathcal{L}$ possesses *property* L' iff each compact subset of G is contained in a compactly generated open subgroup H of G such that G/H is torsion-free.

8.13 Remark Pontryagin [1966, p. 260 (A)] defined a discrete abelian group as having "property L" iff each finite subset of G is contained in a finitely generated subgroup H of G such that G/H is torsion-free. He then proceeded to show (in his Theorem 48) that $G \in \mathcal{L}_d$ has property L iff $\hat{G}$ is locally connected. Much later Fan (1970) generalized property L to the property we call L' in Definition 8.12 and succeeded in generalizing Pontryagin's theorem to include all LCA groups. We now prove Fan's theorem. (Our proof is essentially that of Fan, with some modifications. Fan proved his theorem without invoking Pontryagin's Theorem 48. We will prove and use one direction of Pontryagin's theorem and also take a slight shortcut by making use of Proposition 8.9.)

8.14 *Theorem* A group $G \in \mathcal{L}$ has property L' iff $\hat{G}$ is locally connected.

Proof: We divide the rather lengthy argument into two parts.

(I) Assume that $\hat{G}$ is locally connected.

(a) First consider the case of discrete G. (This is one direction of Pontryagin's Theorem 48.) Given a finite subset S of G we must find a finitely generated subgroup H of G such that $S \subseteq H$ and G/H is torsion-free. To simplify notation set $B = t(G)$. Then $\hat{B} [\cong \hat{G}/A(\hat{G},B)]$ is at once locally connected and totally disconnected, and hence B is finite (cf. the last part of the proof of Proposition 8.10). Let G' be the torsion-free group G/B and let $\pi:G \to G'$ be the natural map. Let S' be the subgroup of G' generated by the finite set $\pi(S)$ and let H' be a pure subgroup of G' such that $H' \supseteq S'$ and $r(H') = r(S')$ [see P.10(a)]. In particular, H' has finite rank, and the purity of H' shows that G'/H' is torsion-free. Now $(G')\hat{} \cong A(\hat{G},B)$, and since B is finite, $A(\hat{G},B)$ is open in the locally connected group $\hat{G}$, whence $(G')\hat{}$ is locally connected too. But then $(H')\hat{}$, being a quotient of $(G')\hat{}$, must also be locally connected. Since H' is a discrete torsion-free group of finite rank it follows from Proposition 8.10 that H' is finitely generated. Letting $H = \pi^{-1}(H')$, it follows from the fact that ker π is the finite subgroup B of G that H is a finitely generated subgroup of G. It is clear that $S \subseteq H$ and that G/H (which is isomorphic to the torsion-free group G'/H') is torsion-free. This completes the proof of case (a).

(b) Now let G be arbitrary. Since $\hat{G}$ is locally connected we use Proposition 8.9 to write $\hat{G} \cong R^n \times K \times D$, where $n \in Z^{+0}$, K is compact, connected, and locally connected, and D is discrete. Hence we can write $G = G_0 \oplus A$, where $G_0 \cong R^n \times \hat{D}$ is an open subgroup of G and $A \cong \hat{K}$ is a discrete subgroup of G. To show that G has property L', let S be any compact subset of G. If $\pi:G \to A$ denotes the projection map then $\pi(S)$ is a compact and hence finite subset of A. But $\hat{A}$ $(\cong K)$ is compact and locally connected, so by (a) there is a finitely generated subgroup H_0 of A such that $\pi(S) \subseteq H_0$ and A/H_0 is torsion-free. Now let H be the open subgroup $G_0 + H_0$ of G. Since G_0 $(\cong R^n \times \hat{D})$ is compactly generated and H_0 is finitely generated, it is clear that H is compactly generated. It is easily checked that $S \subseteq H$ and that G/H (which is isomorphic to A/H_0) is torsion-free. Therefore G has property L', which completes the proof of (I).

(II) Assume conversely that G has property L'. To show that $\hat{G}$

is locally connected let U be a basic neighborhood of 1 in $\hat{G}$ defined by $U = \{\gamma \in \hat{G} : \gamma(S) \subseteq V\}$ for some compact subset S of G and some neighborhood V of 1 in T. All that we need do is find some neighborhood U_0 of 1 in $\hat{G}$ such that U_0 is connected and $U_0 \subseteq U$.

Since G has property L' there is a compactly generated open subgroup H of G such that $S \subseteq H$ and G/H is torsion-free. Now $A(\hat{G},H)[\cong (G/H)\hat{\ }]$ is compact and connected [P.28(b)] and evidently $A(\hat{G},H) \subseteq U$. Since H is compactly generated it follows from P.26(a) and duality that $\hat{H}$ is locally.connected (cf. 8.9), which is to say that $\hat{G}/A(\hat{G},H)$ is locally connected. We may then find a base $\{U_i' : i \in I\}$ at the identity of $\hat{G}/A(\hat{G},H)$ consisting of open connected sets U_i' having compact closure $\bar{U}_i'$ (here I is an index set). Let π be the natural map from $\hat{G}$ onto $\hat{G}/A(\hat{G},H)$. Now since ker π $[= A(\hat{G},H)]$ is compact [P.22(e)], it is evident that $\pi^{-1}(\bar{U}_i')$ is compact in G for each $i \in I$. Now we have $\cap_{i \in I} \pi^{-1}(\bar{U}_i') = A(\hat{G},H) \subseteq U$, whence an elementary topological argument shows that there is some $j \in I$ such that $\pi^{-1}(\bar{U}_j') \subseteq U$ [cf. §13(H) on p. 74 of Pontryagin (1966)]. Set $U_0 = \pi^{-1}(U_j')$. Then U_0 is a neighborhood of 1 in $\hat{G}$ such that $U_0 \subseteq U$. It remains only to show that U_0 is connected. However, this may be easily proved (for instance, by a straightforward modification of §7.14 of [HR]) from the connectedness of $A(\hat{G},H)$ and U_j'. ∎

Having got through this massive proof, let us pause to examine the property L (i.e., L' for discrete groups) more closely. The following lemma is helpful. (The result is purely algebraic, but we cannot resist the temptation to prove one direction of it by topological results already obtained.)

8.15 Lemma Let $G \in \mathcal{L}_d$ be torsion-free. Then G has property L iff each subgroup H of G having finite rank is free abelian.

Proof: Assume that G has L. Then by Theorem 8.14 the compact group $\hat{G}$ is locally connected. If H is a subgroup of G having finite rank, then $\hat{H}$ (being a quotient of $\hat{G}$) is locally connected, whence by 8.10, H is finitely generated. Since H is torsion-free it follows from P.6(b) that H is free abelian.

Conversely, assume that each finite-rank subgroup of G is free abelian and let S be any finite subset of G. Then gp(S) has finite rank, whence by P.10(a) we may find a pure subgroup H of G such that $r(H)$ is finite and gp(S) $\subseteq H$. Thus $S \subseteq H$, H is finitely generated (since free abelian of finite rank) and G/H is torsion-free. Therefore G has property L. ∎

8.16 Remarks (a) By combining the preceding lemma with 8.9 and 8.14 we arrive at another characterization of the locally connected LCA groups [given in Dixmier (1957)]: A group $G \in \mathscr{L}$ is locally connected iff $G \cong R^n \times K \times D$, where $n \in Z^{+0}$, $D \in \mathscr{L}_d$, and K is a compact connected group such that each subgroup of $\hat{K}$ having finite rank is free abelian. As we mentioned earlier, the question of local connectedness of LCA groups centers on the compact connected groups. In (b) through (e) below we examine such groups.

(b) It follows from (a) (taking into account §24.28 of [HR]) that a compact connected group $K \in \mathscr{L}$ is locally connected iff each finite dimensional quotient of K by a closed subgroup is a product of (finitely many) copies of T.

(c) So far we have not seen any examples of compact, connected, and locally connected groups $K \in \mathscr{L}$ which are not of the form T^m for some cardinal $\mathfrak{m}$. If we look only at second-countable (i.e., metrizable, by P.33) compact groups K we never shall. Indeed, if we combine (a) with Pontryagin's theorem given in P.4(d) and take P.19(e) into account, we arrive at the following: A compact, connected, and second-countable group $K \in \mathscr{L}$ is locally connected iff $K \cong T^m$, where $\mathfrak{m} \leq \aleph_0$. This result dates back to the fundamental paper of Pontryagin (1934). It is a special case of Proposition 8.17 below.

(d) Here is an example of a compact, connected, and locally connected LCA group K which does not have the form T^m for any $\mathfrak{m}$. Let S be the (discrete) Specker group of P.4(e) and set $K = \hat{S}$. Then K is evidently compact and connected. Since each countable subgroup of S is free abelian, it is evident that each finite-rank subgroup of S is free abelian. Therefore by (a), K is locally connected. But since S is not free abelian, K cannot have the form T^m for any cardinal $\mathfrak{m}$.

(e) The structure of compact, connected, and locally connected groups K seems in general quite complicated. In view of (a) (or of 8.14 and 8.15), the problem of describing such groups can be thought of as a purely algebraic problem. An abelian group A with the property that each of its countable subgroups is free abelian is called $\aleph_1$-*free* [cf. p. 94 of Fuchs (1970)]. It is easily shown [using Pontryagin's theorem mentioned in P.4(d)] that $A \in \mathscr{A}$ is $\aleph_1$-free iff each subgroup of A having finite rank is free abelian. We can therefore say that a compact connected group $K \in \mathscr{L}$ is locally connected iff $\hat{K}$ is $\aleph_1$-free. There is an abundance of $\aleph_1$-free groups, but their general structure seems to be unknown.

Part (c) of the above at least enables us to give a complete description of the metrizable locally connected LCA groups. The next result is a slight generalization of Theorem 49 of Pontryagin (1966).

8.17 Proposition Let $G \in \mathcal{L}$ be metrizable. Then G is locally connected iff $G \cong R^n \times T^m \times D$, where $n \in Z^{+0}$, $m \leq \aleph_0$ and $D \in \mathcal{L}_d$. If in addition G is compact, then G is locally connected iff $G \cong T^m \times F$, where $m \leq \aleph_0$ and F is finite.

Proof: The first part of the proposition follows immediately from 8.9 and 8.16(c), while the second part is an obvious consequence of the first. ∎

Having rendered, as we hope, sufficient justice to the locally connected LCA groups, we now acquaint ourselves with the arcwise connected LCA groups. By way of review, and to establish terminology, we recall that a topological space X is arcwise connected iff for each pair of points x_0 and x_1 of X there is a continuous function $f:[0, 1] \to X$ such that $f(0) = x_0$ and $f(1) = x_1$. Such a function f is called a path from x_0 to x_1. (We follow [HR] in this definition of arcwise connectedness. Many authors, of course, have a more stringent definition in mind when referring to arcwise connectedness, using the term "path-connected" for the property of our definition. Since the distinction evaporates in Hausdorff spaces, we shall run into no ambiguity.)   For a fixed $x_0 \in X$, the set of all $x \in X$ for which there is a path from x_0 to x is called the arcwise connected component (or path-component) of x_0. We will show that the arcwise connected component of 0 in any LCA group G is just the union of the one-parameter subgroups of G. This important and by no means obvious fact will enable us finally to obtain a dual characterization of arcwise connectedness. In this connection we shall find it convenient to cite a few results about the Ext functor in abelian groups which, because of their special nature and because we require them only here, we have not thought worthwhile to set forth in the Preliminaries chapter. Full references to Fuchs (1970) will, however, be given. The connection between arcwise connectedness and the Ext functor constitutes one of the most fascinating examples of the interplay between topology and algebra of which the author is aware. And the whole subject of arcwise connected LCA groups has a rather enigmatic ingredient [see 8.25(d)].

The groups R and T are evidently arcwise connected, and since a product of arcwise connected spaces is arcwise connected, so is $R^n \times T^m$, for any $n \in Z^{+0}$ and cardinal m. It will be seen, on the other hand [see 8.22(b) and 8.25(b)], that $\hat{Q}$ and the group $\hat{S}$ of 8.16(d) are not arcwise connected. Readers having trouble finding examples of arcwise connected LCA groups not of the form $R^n \times T^m$ will soon see a (perhaps unexpected) reason for their difficulty. We begin with a moderately involved, but necessary, technical result from Dixmier (1957).

8.18 Lemma Let $G \in \mathscr{L}$ and let I be the closed interval $[0,1]$. Let $t \to \gamma_t$ be a continuous mapping from I into $\hat{G}$ satisfying $\gamma_0 = 1$. Then there exists a mapping $t \to f_t$ from I into $\mathrm{Hom}(G, R)$ such that (a) for each $x \in G$, $f_t(x)$ is continuous in t, (b) $\gamma_t = \exp(2\pi i f_t)$ for each $t \in I$, and (c) $f_0 \equiv 0$.

Proof: If ϕ is any continuous function from I into T satisfying $\phi(0) = 1$, we may construct a unique continuous function $\psi : I \to R$ satisfying $\phi = \exp(2\pi i \psi)$ and $\psi(0) = 0$. [This is, of course, a very special case of a rather general construction; see, for example, K. Kuratowski, *Set Theory and Topology*, Pergamon Press, 1972 (§§3–5 of Chapter XXI).] If we fix $x \in G$ and set $\phi(t) = \gamma_t(x)$, our hypotheses imply that ϕ is a continuous function from I into T such that $\phi(0) = 1$. The corresponding ψ found above will then be a function $f_t(x)$ which is continuous in t and satisfies $\gamma_t(x) = \exp(2\pi i f_t(x))$ for $0 \leq t \leq 1$, and $f_0(x) = 0$. If for a given t we let x vary we obtain a function $f_t : G \to R$. We claim that the mapping $t \to f_t$ is the one sought. It is evident that conditions (a), (b), and (c) hold. The less obvious part is that $f_t \in \mathrm{Hom}(G, R)$ for each $t \in I$: We must show that for each $t \in I$, (A) f_t is a homomorphism and (B) f_t is continuous.

(A) We must show that for any $t \in I$ and $x, y \in G$ we have $f_t(x + y) = f_t(x) + f_t(y)$. Fixing x and y, consider the function $h(t) = f_t(x + y) - f_t(x) - f_t(y)$. We know by (a) that $h : I \to R$ is continuous. Since $\gamma_t(x + y) = \gamma_t(x) \cdot \gamma_t(y)$ we have by (b) that $\exp(2\pi i h(t)) = 1$ for all $t \in I$, so h takes on only integer values. But by (c), $h(0) = 0$, so by the intermediate value theorem, $h(t) = 0$ for all $t \in I$. Hence $f_t(x + y) = f_t(x) + f_t(y)$, as desired.

(B) Let us first observe that if this lemma holds for two LCA groups G_1 and G_2 then it also holds for $G_1 \times G_2$. We then see from P.29 that we need only prove the lemma [and hence need only prove part (B)] for two cases: (i) for $G = R$ and (ii) for G containing a compact open subgroup. Let us take the easier case (i) first. If for fixed $t \in I$ we have $\exp(2\pi i f_t) = \gamma_t \in \hat{R}$ then by P.20(b) we see that there exists $r \in R$ such that $\exp(2\pi i f_t(x)) = \exp(2\pi i r x)$ for all $x \in R$. We conclude that $f_t(x) = rx + M(x)$ where $M(x)$ is an integer for all $x \in R$. Since by (A), f_t is a homomorphism, we conclude that $M : R \to Z$ is also a homomorphism. Since R is divisible and Z is reduced, we have $M \equiv 0$, so $f_t(x)$ is the continuous function rx of x. This proves case (i). For case (ii) let K be a compact open subgroup of G, and let γ'_t be the restriction of γ_t to K. The mapping $t \rightarrow \gamma'_t$ is then continuous from I into $\hat{K}$. But $\hat{K}$ is discrete, I is connected, and $\gamma'_0 = 1$, whence we conclude that $\gamma'_t = 1$ for all $t \in I$. If we can show that for each $t \in I$ the homomorphism f_t vanishes on the open subgroup K, then f_t [being a homomorphism, by (A)] must be continuous. To this end let $x \in K$. Since $\exp(2\pi i f_t(x)) = \gamma'_t(x) = 1$, we see that $f_t(x)$ is an integer for each $t \in I$. However, $f_t(x)$ is continuous in t and $f_0(x) = 0$, whence $f_t(x) = 0$ for all $t \in I$. Since $x \in K$ was arbitrary, we are done with case (ii) and hence with the proof of the lemma. ∎

We can now prove a result from Dixmier (1957) which is fundamental to our study of arcwise connectedness.

8.19 Theorem For any $G \in \mathcal{L}$ the arcwise connected component A of 0 is the union U of the one-parameter subgroups of G. In particular, G is arcwise connected iff each element of G lies on some one-parameter subgroup of G.

Proof: It is immediately verified that $U \subseteq A$. To show the opposite inclusion, it will be convenient to think of G as $\hat{H}$, where $H = \hat{G}$. Pick any $\gamma \in A$. Then there is a path in $\hat{H}$ from 1 to γ. More precisely, let the path be the continuous mapping $t \rightarrow \gamma_t$ from $I = [0,1]$ to $\hat{H}$ with $\gamma_0 = 1$ and $\gamma_1 = \gamma$. Let $t \rightarrow f_t$ be the map from I to $\text{Hom}(H,R)$ guaranteed to exist in the previous lemma. For our purposes it will suffice to look only at $f_1 \in \text{Hom}(H,R)$. For any fixed $r \in R$ define a function $\phi_r : H \rightarrow T$ by the rule $\phi_r(x) = \exp(2\pi i r f_1(x))$ for all $x \in H$. It is evident that $\phi_r \in \hat{H}$ for each $r \in R$, and it is easily verified that the mapping $\phi : R \rightarrow \hat{H}$ defined by $\phi(r) = \phi_r$ for each $r \in R$ is a continuous homomorphism. Now our original γ is just $\gamma_1 = \exp(2\pi i f_1) = \phi_1 = \phi(1)$. Since $\phi \in \text{Hom}(R, \hat{H})$ we see that γ lies on the one-parameter subgroup $\phi(R)$ of $\hat{H}$. Since $\gamma \in A$ was arbitrary, we have $A \subseteq U$, which completes the proof. ∎

8.20 Remark It is readily verified that the arcwise connected component A of 0 of any $G \in \mathcal{L}$ is actually a subgroup of G [cf. 4.13(d)]. It follows moreover from Theorem 8.19 that A is dense in G iff $G \in \mathcal{S}^*(R)$. We may then say, by 4.8, that $G \in \mathcal{L}$ is connected iff the arcwise connected component A of 0 is dense in G. But in "most" connected groups A is actually a proper subgroup of G, as we shall see below.

Now that we have Theorem 8.19 it would be handy to have a criterion for determining whether or not every element of a given group $G \in \mathscr{L}$ actually lies on some one-parameter subgroup. The following result from Dixmier (1957) is just the thing.

8.21 Lemma The following are equivalent for any $G \in \mathscr{L}$:
(a) G is the union of its one-parameter subgroups.
(b) Each continuous character on $\hat{G}$ has the form $\exp(if)$ for some $f \in \mathrm{Hom}(\hat{G}, R)$.

Proof: Part (b) is by P.20(b) clearly equivalent to the condition that each $\rho \in \mathrm{Hom}(\hat{G}, T)$ has the form $\sigma \circ \tau$ for some $\tau \in \mathrm{Hom}(\hat{G}, R)$ and $\sigma \in \mathrm{Hom}(R, T)$. Hence the result is a special case of 4.31. (This special case is also proved in §24.43 of [HR].) ∎

8.22 Examples (a) Every metrizable connected and locally connected LCA group G is arcwise connected. Indeed, it follows from 8.17 that G must have the form $R^n \times T^m$ and is hence arcwise connected.
(b) The group $\hat{Q}$ is not arcwise connected. Indeed, it is not hard to construct $\gamma \in \mathrm{Hom}(Q, T)$ which does not have the form $\exp(if)$ for any $f \in \mathrm{Hom}(Q, R)$ [see §25.26(d) of [HR]], whence by Lemma 8.21 and Theorem 8.19, $\hat{Q}$ is not locally connected. [We can also argue as follows: $\hat{Q}$ is not locally connected (8.11(a)), whence by 8.27 (proved independently below), $\hat{Q}$ cannot be arcwise connected.]

We are now going to see just how hard it is for a compact connected LCA group to be arcwise connected. At this point a very interesting class of abelian groups enters the picture. An abelian group A is called a *Whitehead group* iff $\mathrm{Ext}(A, Z) = \{0\}$. [See Chapter IX of Fuchs (1970) for a definition and discussion of Ext.]

8.23 Remark A discussion of Whitehead groups may be found in §99 of Fuchs (1973). We cite the following facts therefrom:
(a) Free abelian groups are Whitehead groups.
(b) If A is a Whitehead group and B is a subgroup of A such that $|B| \leq \aleph_0$, then B is free abelian. In particular, A is torsion-free. [From 8.16(e) and 8.15 we conclude that Whitehead groups satisfy property L.]
(c) The Specker group S is not a Whitehead group. [This is a special case of Proposition 99.2 of Fuchs (1973).]

We now prove a result [from Dixmier (1957)] which, in light of Lemma 8.21, will indicate the close relationship between "Whiteheadness" and arcwise connectedness.

8.24 Lemma The following are equivalent for any $A \in \mathscr{L}_d$:
(a) Each $\gamma \in \hat{A}$ has the form $\exp(if)$ for some $f \in \mathrm{Hom}(A, R)$.
(b) A is a Whitehead group.

Proof: We make use of some elementary properties of the functors Hom and Ext. We begin with the exact sequence $0 \to Z \to R \to T \to 0$, where the maps indicated by the arrows are the obvious ones. (We shorten $\{0\}$ to 0.) By applying Theorem 51.3 of Fuchs (1970) we obtain the following long exact sequence, where the maps are as indicated in the reference cited: $0 \to \mathrm{Hom}(A, Z) \to \mathrm{Hom}(A, R) \to \mathrm{Hom}(A, T) \to \mathrm{Ext}(A, Z) \to \mathrm{Ext}(A, R) \to \mathrm{Ext}(A, T) \to 0$. Now R and T are divisible, whence by (B) on p. 222 of Fuchs (1970) we have $\mathrm{Ext}(A, R) = \mathrm{Ext}(A, T) = 0$. Our long exact sequence now reduces to the exact sequence $0 \to \mathrm{Hom}(A, Z) \to \mathrm{Hom}(A, R) \to \mathrm{Hom}(A, T) \to \mathrm{Ext}(A, T) \to 0$. Now condition (a) is equivalent to the surjectivity of the map from $\mathrm{Hom}(A, R)$ to $\mathrm{Hom}(A, T)$ [see Fuchs (1970, §44)]. However, the exactness of the sequence implies that this map is surjective iff $\mathrm{Ext}(A, Z) = 0$, i.e., iff A is a Whitehead group. This proves the equivalence of (a) and (b). ∎

8.25 Remarks (a) By combining the previous lemma with Lemma 8.21 and Theorem 8.19 we conclude the following (Dixmier 1957): A compact group $G \in \mathscr{L}$ is arcwise connected iff $\hat{G}$ is a Whitehead group.

(b) Using the result (a), Dixmier (1957) first gave an example of a compact, connected, and locally connected LCA group G which is not arcwise connected. In fact, this G is none other than the group $\hat{S}$ of 8.16(d). [The fact that $\hat{S}$ is not arcwise connected follows from (a) and 8.23(c). Our glib invocation of 8.23(c) conceals a considerable amount of computation, for which the reader may consult Dixmier's paper.]

(c) Whitehead (1956) first proved the following fact: A compact second countable group $G \in \mathscr{L}$ is the union of its one-parameter subgroups iff G is a (countable) product of copies of T. [This follows immediately from 8.21, 8.24, P.19(e), and 8.23(b).] Taking P.33 and Theorem 8.19 into account, we arrive at the following fact: A compact metrizable group $G \in \mathscr{L}$ is arcwise connected iff $G \cong T^{\mathfrak{m}}$ for some $\mathfrak{m} \leqq \aleph_0$. (This is a special case of Theorem 8.27 below.)

(d) Is there a compact arcwise connected group $G \in \mathscr{L}$ which is not a product of copies of T? By (a) the question comes to: Is there a Whitehead group which is not free abelian? For many years this was an open and much-discussed problem of abelian group theory, until Shelah (1974) showed that the question is undecidable in ordinary set theory (the Zermelo-Fraenkel axioms together with the axiom of choice). It seems strange that such a concrete-sounding question as "Is every compact arcwise connected LCA group a torus of some dimension?" should be left with such an answer.

Let us close by bringing together the more positive aspects of our findings by proving two theorems from Dixmier (1957). The second of these theorems indicates the relation between local connectedness and arcwise connectedness in LCA groups.

8.26 Theorem A group $G \in \mathscr{L}$ is arcwise connected iff $G \cong R^n \times K$, where $n \in Z^{+0}$ and K is a compact abelian group whose dual $\hat{K}$ is a Whitehead group.

Proof: This follows immediately from P.27(e) and 8.25(a). ∎

8.27 Theorem Consider the following properties for $G \in \mathscr{L}$:
(a) $G \cong R^n \times T^m$ for some $n \in Z^{+0}$ and cardinal number m.
(b) G is arcwise connected.
(c) G is connected and locally connected.
Then we have:
(i) (a) $\Rightarrow$ (b) $\Rightarrow$ (c).
(ii) (a), (b), and (c) are equivalent if G is metrizable.
(iii) (c) $\not\Rightarrow$ (b) generally.

Proof: We have already pointed out (in the paragraph before Lemma 8.18) that (a) $\Rightarrow$ (b). If (b) holds, then G is evidently connected. Moreover by Theorem 8.26, G has the form $R^n \times K$, where $\hat{K}$ is a discrete Whitehead group. Then from 8.23(b) we know that $\hat{K}$ has property L (hence L'), so K is locally connected by Theorem 8.14. Thus G is locally connected and we have (b) $\Rightarrow$ (c). This proves (i). If G is metrizable and (c) holds, then Proposition 8.17 shows that (a) holds, whence we have (ii). Finally, (iii) follows from 8.25(b). ∎

Miscellanea

8.28 (Cf. 8.3.) For any $G \in \mathscr{L}$, the Bohr compactification $\beta(G)$ of G is connected iff $|\text{im } \gamma| \geq \aleph_0$ for each $\gamma \neq 1$ in $\hat{G}$.

8.29 A nontrivial group $G \in \mathscr{L}$ is connected iff there exists $\gamma \in \hat{G}$ such that ker γ is a proper subgroup of $c(G)$.

8.30 Let $G \in \mathscr{L}$ be connected and let γ_1 and γ_2 be in $\hat{G}$. Then we have:
(a) If ker $\gamma_1 \subseteq$ ker γ_2 there exists $n \in Z$ such that $\gamma_2 = \gamma_1^n$.
(b) If ker $\gamma_1 =$ ker γ_2 then either $\gamma_2 = \gamma_1$ or $\gamma_2 = \gamma_1^{-1}$.

8.31 In 8.2 it was shown that T "decides" the connectedness of a group $G \in \mathscr{L}$ in the sense that G is connected iff each $f \neq 0$ in $\text{Hom}(G,T)$ is surjective. In fact, T is the only (not totally disconnected, of course) LCA group H with the property that $G \in \mathscr{L}$ is connected iff each $f \neq 0$ in $\text{Hom}(G,H)$ is surjective.

8.32 (Armacost 1972) Let $G \in \mathscr{L}$ and let N be the set of nonsurjective $\gamma \in \hat{G}$. The following are equivalent:
(a) N is a subgroup of $\hat{G}$.
(b) N is a closed subgroup of $\hat{G}$.
(c) Each surjective $\gamma \in \hat{G}$ is an open mapping.

8.33 Let $G \in \mathscr{L}$ be compact and connected. If G is locally connected, then $t(G)$ is dense in G, but not conversely.

8.34 Let $G \in \mathscr{L}$ be compact and connected. Then G is locally connected iff for each finite subset $\{\gamma_1, \ldots, \gamma_n\}$ of $\hat{G}$ there is a compact connected subgroup K of G such that $K \subseteq \Pi_{i=1}^n$ ker γ_i and G/K is a product of finitely many copies of T.

8.35 Let $G \in \mathscr{L}$ have the property that G/H is locally connected for each closed subgroup $H \neq \{0\}$ of G. Then either $G \cong F_p$, $G \cong J_p$ (for some $p \in \mathscr{P}$), or else G is itself locally connected. [Since discrete groups are locally connected, assume that G is not discrete. If G is totally disconnected use 1.6. If $c(G) \neq \{0\}$ use 6.9 and P.27(e) to reduce the problem to the case of compact connected G. By 8.16(e) each proper subgroup of $\hat{G}$ is $\aleph_1$-free. Then show that $\hat{G}$ itself must be $\aleph_1$-free.]

8.36 What conclusions can be drawn if in 8.35 we replace "locally connected" by "arcwise connected"?

8.37 Let $G \in \mathscr{L}$. Then $G \cong R^n$ for some $n \in Z^{+0}$ iff the following two conditions hold:
(a) $\hat{G}$ is arcwise connected.
(b) Each $f \neq 0$ in $\text{Hom}(G, R)$ is surjective.

8.38 (Dixmier 1957) A group $G \in \mathscr{L}$ is locally arcwise connected (by which we mean, of course, that G has a basis consisting of open arcwise connected subsets) iff $G \cong R^n \times K \times D$, where $n \in Z^{+0}$, $D \in \mathscr{L}_d$, and K is a compact connected group such that (a) $\hat{K}$ is a Whitehead group and (b) each pure subgroup of $\hat{K}$ having finite rank is a direct summand of $\hat{K}$. [It is relevant to point out that Rotman proved, a few years after the appearance of Dixmier's paper, that condition (b) already is implied by condition (a), for which see Theorem 99.1 of Fuchs (1973). It then follows from Dixmier's result and Theorem 8.26 that an arcwise connected group must be locally arcwise connected. This would of course be evident if we knew that a Whitehead group must be free abelian. But we don't.]

8.39 Let $G \in \mathscr{L}$ be compact. Regarding G as a topological group with base point 0 we can consider the homotopy groups $\pi_n(G)$. Enochs (1964) shows that for $n \geq 2$ we have $\pi_n(G) = \{0\}$, while $\pi_1(G) \simeq \text{Hom}(\hat{G}, Z)$. Now if A is a nontrivial Whitehead group, then $\text{Hom}(A, Z) \neq \{0\}$. [This follows from §99.1 or §99.4 of Fuchs (1973).] Since by definition G is simply connected iff G is arcwise connected and satisfies $\pi_1(G) = \{0\}$ we conclude from 8.25(a) and Enochs' result that no nontrivial compact LCA group is simply connected.

9

More Splitting, and Some Homological Methods

"Bring me my arrows . . . " [William
Blake, *Milton*]

In this chapter we pursue the subject of splitting in LCA groups. We begin
by proving a structure theorem of Robertson which says, roughly, that
copies of Q and $\hat{Q}$ may be split from any LCA group in such a way that, in a
sense, no copies remain in what is left over. As far as the author is aware,
Robertson's theorem constitutes the most refined structure theorem known
for arbitrary LCA groups. Having made some deductions from this
theorem, we pass to complete our discussion of groups with the "universal"
splitting property of Theorem 6.16. These groups turn out to be the
"injectives" of the category $\mathscr{L}$. In order to study these groups (and their
duals, the "projectives") we introduce the rudiments of homological
algebra in $\mathscr{L}$. A brief sketch of further homological methods and problems
will be found in 9.27.

The fundamental structure theorem P.29 enables us to write any LCA
group in the form $R^n \times G_0$, where G_0 has a compact open subgroup, i.e.,
$c(G_0)$ is compact. It is evident that G_0 can contain no copy of R. By 6.18
we know further that any copy of Q^{m^*} which G_0 might contain as a closed
subgroup must split from G_0. It is then reasonable to ask whether we can
split a "largest" copy of Q^{m^*} from G_0, that is, whether we can write
$G_0 \cong Q^{m^*} \times H$, where H contains no copy of Q. It turns out that we can do
this and more. The precise result is the theorem of Robertson mentioned
above. We first prove a lemma. Recall [P.9(f)] that every abelian group G
has a unique maximal divisible subgroup $d(G)$.

9.1 Lemma Let $G \in \mathcal{L}$ have compact identity component. Then $G \cong Q^{m^*} \times H$, where m is some cardinal number and $H \in \mathcal{L}$ satisfies $d(H) \subseteq b(H)$.

Proof: Set $A = \overline{d(G)} \subseteq G$. Then A is a densely divisible LCA group, whence by 6.14 we can write $A = b(A) \oplus B$, where $B \cong Q^{m^*}$ for some cardinal m. Since B is divisible and $B \cap b(G) = \{0\}$ it follows from P.9(d) that there is a subgroup H of G such that $H \supseteq b(G)$ and $G = B + H$. Now $b(G)$ is open in G [P.17(c)], so H is open too, whence by 6.8 we have $G = B \oplus H$. We will be done if we show that $d(H) \subseteq b(H)$. To this end, let h be any element of $d(H)$. Since $d(H) \subseteq A$ and $A = b(A) \oplus B$ we can find $x \in b(A)$ and $y \in B$ such that $h = x + y$. But $b(A) \subseteq b(G) \subseteq H$, so $y = h - x \in H$. Thus $y \in B \cap H = \{0\}$, so $h = x$ is compact, i.e., $h \in b(H)$. Therefore we have $d(H) \subseteq b(H)$, and the proof is complete. ∎

It is evident that H in Lemma 9.1 can contain no copy of Q, since such a copy would be contained in $d(H) \subseteq b(H)$, whereas $b(Q) = \{0\}$.

The following definition was introduced in Armacost and Armacost (1978).

9.2 Definition A group $G \in \mathcal{L}$ is said to be *residual* iff $d(G) \subseteq b(G)$ and $d(\hat{G}) \subseteq b(\hat{G})$.

We note that a residual group cannot have Q as a closed subgroup or $\hat{Q}$ as a quotient by a closed subgroup. A residual group G may, however, contain a copy of $\hat{Q}$ as a closed subgroup (e.g., take $G = T^{\aleph_0}$; cf. 6.19). Note that a topological torsion group G is trivially residual, since $b(G) = G$ and $b(\hat{G}) = \hat{G}$ (cf. 3.5). Indeed, the class of residual groups is quite extensive. This fact (and the reason for the name "residual") will be seen from Robertson's structure theorem, which we now present. Our proof [from Armacost and Armacost (1978)] is essentially that given by Robertson in his unpublished paper (1968).

9.3 Theorem Any group $G \in \mathcal{L}$ has the form $R^n \times Q^{m^*} \times (\hat{Q})^n \times E$, where $n \in Z^{+0}$, m and n are cardinal numbers, and E is residual.

Proof: Use P.29 to write $G \cong R^n \times G_0$, where $c(G_0)$ is compact. By 9.1 we have $G_0 \cong Q^{m^*} \times H$, where m is some cardinal and $d(H) \subseteq b(H)$. Now since $c(H)$ is compact, $c(\hat{H})$ is necessarily compact too, so we may apply 9.1 to $\hat{H}$ to obtain $\hat{H} \cong Q^{n^*} \times K$, where n is some cardinal and $d(K) \subseteq b(K)$. Then setting $E = \hat{K}$ we have $G \cong R^n \times$

$Q^{m^*} \times (\hat{Q})^n \times E$, and it only remains to show that E is residual. We already know that $d(\hat{E}) \subseteq b(\hat{E})$. Moreover, every element of $d(H)$ is compact, and since $H \cong (\hat{Q})^n \times \hat{K}$, it is evident that every element of $d(\hat{K})$ is compact. Thus $d(\hat{K}) \subseteq b(\hat{K})$, or $d(E) \subseteq b(E)$. Thus E is residual. ∎

9.4 Remarks (a) It follows from Robertson's theorem that in studying the indecomposable LCA groups (cf. 6.24 through 6.27) we may restrict our attention to residual groups.

(b) It may be shown [see Armacost and Armacost (1978)] that the cardinal numbers n, m, and n are uniquely determined by G and that the "residual part" E of G is unique up to topological isomorphism. One consequence of this is that $G \in \mathcal{L}$ is self-dual (cf. 4.37) iff the residual part of G is self-dual.

We now apply Theorem 9.3 to study densely divisible groups. The theorem may be used to yield an immediate strengthening of 6.14. Leaving the reader to formulate this, we make the additional assumption that G is torsion-free.

9.5 Proposition Let $G \in \mathcal{L}$ be torsion-free. Then G is densely divisible iff $G \cong R^n \times Q^{m^*} \times (\hat{Q})^n \times E$, where $n \in Z^{+0}$, m, and n are cardinals, and E is a densely divisible torsion-free topological torsion group.

Proof: Assuming that G is densely divisible and applying Theorem 9.3, we evidently need only show that E is a topological torsion group. Since $d(E) \subseteq b(E)$ and $d(E)$ is dense in E, we have $b(E) = E$. But E is also torsion-free, whence $\hat{E}$ is densely divisible (4.15), and it follows in the same way that $b(\hat{E}) = \hat{E}$. By P.22(g) both E and $\hat{E}$ are totally disconnected, so by 3.5, E is a topological torsion group. The converse is evident. ∎

Mackey (1946) proved an important structure theorem for divisible torsion-free LCA groups. A proof also appears in §25.33 of [HR]. We present the result here as a consequence of the preceding proposition.

9.6 Corollary Let $G \in \mathcal{L}$ be divisible and torsion-free. Then $G \cong R^n \times Q^{m^*} \times (\hat{Q})^n \times E$, where $n \in Z^{+0}$, m and n are cardinals, and E is the minimal divisible extension of a product of groups J_p for various primes p.

Proof: Upon applying 9.5 to G we see that only the part about E needs verifying. By 3.5, E has a compact open subgroup K such that E/K is a torsion group. Since E is divisible and torsion-free, it follows from P.31(c) that E is a minimal divisible extension of K. Finally, since K is compact, totally disconnected, and torsion-free, it follows from P.24(a) that K is a product of various J_p-groups. ∎

For an interesting generalization of 9.6 see Corwin (1976).

Before taking up the universal splitting groups, we will need two general structural results, which have considerable interest in their own right. It is well known that any compact group $G \in \mathscr{L}$ can be embedded as a closed subgroup of T^m for some cardinal m. Indeed, if G is an arbitrary LCA group and T_γ denotes a copy of T for each $\gamma \in \hat{G}$, it is evident that the map $\phi : G \to \Pi_{\gamma \in \hat{G}} T_\gamma$, defined by $\phi(x) = (\gamma(x))$ for each $x \in G$, is a continuous monomorphism. If in addition G is compact, ϕ is clearly a topological isomorphism onto its image. To extend this embedding theorem to arbitrary LCA groups, we make the following definition from Ahern and Jewett (1965).

9.7 Definition A group $G \in \mathscr{L}$ is called a *d-group* iff $G \cong R^n \times T^m \times D$, where $n \in Z^{+0}$, m is a cardinal number, and D is a discrete divisible group.

The next two results about d-groups are from Ahern and Jewett (1965).

9.8 Proposition Any group $G \in \mathscr{L}$ can be embedded as a closed subgroup of a d-group.

Proof: By P.29 we may assume that G has a compact open subgroup K. By P.9(g) any discrete abelian group can be embedded in some divisible group $D \in \mathscr{L}_d$. Since G/K is discrete, we have now reduced the problem to showing that G can be embedded as a closed subgroup of $T^m \times G/K$ for some cardinal number m. As above, let ϕ be any continuous monomorphism from G into a group T^m and let $\pi : G \to G/K$ be the natural map. Define $\psi : G \to T^m \times G/K$ by $\psi(x) = (\phi(x), \pi(x))$ for all $x \in G$. It is clear that ψ is a continuous monomorphism, so we are left with showing that ψ is an open mapping onto its image. We do this by showing that ψ^{-1} is continuous at the zero of $\psi(G)$. Indeed, let $\{(\phi(x_i), \pi(x_i))\}_{i \in I}$ be a net in $\psi(G)$ converging to zero (here I is some directed set). We must show that $x_i \to 0$ in G. Since G/K is discrete and $\pi(x_i) \to 0 + K$, there exists $i_0 \in I$ such that $x_i \in K$ for $i \geqq i_0$. If ϕ_0 denotes the restriction of ϕ to K, the compactness of K shows that ϕ_0 is a topological isomorphism from K onto $\phi(K)$. But since $\phi(x_i) \to 0$ and $\phi(x_i) = \phi_0(x_i)$ for $i \geqq i_0$, we have $x_i \to 0$ in K and hence in G. ∎

The product of finitely many d-groups is again a d-group. Moreover, d-groups are clearly divisible (perhaps accounting for the name), whence quotients of d-groups are divisible. But we can say more.

9.9 Proposition   A quotient of a d-group by a closed subgroup is again a d-group.

Proof: We prove the equivalent assertion that if f is a continuous open homomorphism from a d-group $R^n \times T^m \times D$ onto some group $G \in \mathscr{L}$, then G is a d-group. Now $H = f(R^n \times T^m \times \{0\})$ is an open divisible subgroup of G, whence by 6.9 we have $G = H \oplus A$, where A is some discrete subgroup of G. Since A must be divisible, it remains only to show that H is a d-group. To this end, let H' be the compact subgroup $f(\{0\} \times T^m \times \{0\})$ of H. Since H' is a quotient of T^m it follows from P.4(d) and duality that $H' \cong T^n$ for some cardinal number $\mathfrak{n}$. By 6.16 we have $H = H' \oplus K$ for some closed subgroup K of H. Since $K \cong H/H'$, it evidently suffices now to show that H/H' is a d-group. Let π be the natural map from H onto H/H'. Now for any element $(\bar{r}, \bar{t}, 0)$ in $R^n \times T^m \times \{0\}$ we have $\pi(f(\bar{r}, \bar{t}, 0)) = \pi(f(\bar{r}, 0, 0) + f(0, \bar{t}, 0)) = \pi(f(\bar{r}, 0, 0))$, since $f(0, \bar{t}, 0) \in H' = \ker \pi$. Thus the continuous homomorphism $\pi \circ f$ maps $R^n \times \{0\} \times \{0\}$ onto H/H'. Since $\pi \circ f$ is open [P.30(b)], we see that H/H' is a quotient of R^n by a closed subgroup. It then follows from P.18(b) and duality that H/H' has the form $R^a \times T^b$, i.e., H/H' is a d-group, as desired. ∎

Now we make use of d-groups.

9.10 Definition   A group $G \in \mathscr{L}$ is called a *universal splitting group* iff G is splitting in the class $\mathscr{L}$ (cf. Definition 6.15).

This terminology is borrowed from Robertson (1968). [Ahern and Jewett (1965) use the term "universal internal direct factor."] We now proceed to identify the universal splitting groups. Our identification of these groups (Theorem 9.12 below) is from Ahern and Jewett (1965). The result was proved independently later by Robertson (1968), based on Dixmier's characterization of the "injective" groups in $\mathscr{L}$ (for which see 9.17 below). We need a preliminary fact.

9.11 Lemma   Let $A \in \mathscr{L}$ be a universal splitting group. If B is a closed subgroup of A which splits from A, then B is also a universal splitting group.

Proof: Let $G \in \mathscr{L}$ have a closed subgroup $H \cong B$. We must show that H splits from G. Now we can write $A \cong B \times C$ for some $C \in \mathscr{L}$. Moreover, $H \times C$ is a subgroup of $G \times C$ in the obvious way, and $H \times C \cong A$. Since A is a universal splitting group, $H \times C$ must split from $G \times C$, so there exists $f \in \text{Hom}(G \times C, H \times C)$ such that f is the identity on $H \times C$. If $i : G \to G \times C$ is the injection and $\pi : H \times C \to H$ the projection, we define $g \in \text{Hom}(G, H)$ by $g = \pi \circ f \circ i$. It is clear that $g(h) = h$ for all $h \in H$, so H splits from G by 6.6. ∎

9.12 Theorem A group $A \in \mathscr{L}$ is a universal splitting group iff $A \cong R^n \times T^m$ for some $n \in Z^{+0}$ and cardinal number $\mathfrak{m}$.

Proof: We have already shown (6.16) that groups of the form $R^n \times T^m$ are universal splitting groups. Conversely, suppose that A is a universal splitting group. By 9.8, A is topologically isomorphic to a closed subgroup H of some d-group. Since H splits from G, then H is a quotient of G, whence by 9.9, H (and therefore A) is a d-group, i.e., A has the form $R^n \times T^m \times D$, where D is discrete and divisible. But neither $Z(p^\infty)$ nor Q is a universal splitting group (see 6.4 and 6.17 respectively). It then follows from P.9(e) and 9.11 that D must be trivial. Thus $A \cong R^n \times T^m$, and the proof is complete. ∎

We now take up some homological questions closely related to Theorem 9.12. $\mathscr{L}$ may be thought of as a category whose objects are the LCA groups and whose morphisms are the continuous homomorphisms between them. The usual machinery of abelian categories does not, however, apply directly to $\mathscr{L}$. [See Moskowitz (1967), from which paper much of what follows is drawn.] It happens that one should frequently restrict one's attention to proper homomorphisms (see P.15) to smooth things along. For example, consider a short exact sequence $0 \to G_1 \overset{f}{\to} G_2 \overset{g}{\to} G_3 \to 0$, where G_1, G_2, G_3 are in $\mathscr{L}$ and f and g are continuous homomorphisms. Let us form the "dual" sequence $0 \to \hat{G}_3 \overset{g^*}{\to} \hat{G}_2 \overset{f^*}{\to} \hat{G}_1 \to 0$. It would be very pleasant if this dual sequence always had to be exact, but such need not be the case. Consider the sequence $0 \to Z \overset{f}{\to} R_d \overset{g}{\to} T \to 0$, where $f(n) = n$ for all $n \in Z$ and $g(r) = \exp(2\pi i r)$ for all $r \in R_d$. It is clear that f and g are continuous homomorphisms and that the sequence is exact. However, the dual sequence $0 \to \hat{T} \overset{g^*}{\to} (R_d)\hat{} \overset{f^*}{\to} \hat{Z} \to 0$ cannot be exact, since im g^* (being countably infinite) cannot be closed in the compact group $(R_d)\hat{}$, so im $g^* = \ker f^*$ is impossible. We are thus led to consider more special short exact sequences.

9.13 Definition Let G_1, G_2, and G_3 be in $\mathscr{L}$ and let $f \in \text{Hom}(G_1, G_2)$ and

$g \in \text{Hom}(G_2, G_3)$. The sequence $0 \to G_1 \xrightarrow{f} G_2 \xrightarrow{g} G_3 \to 0$ is said to be a *proper short exact* (PSE) sequence in $\mathscr{L}$ iff it is exact in the algebraic sense (Fuchs 1970, p. 8) and the maps f and g are proper.

Note that if H is a closed subgroup of $G \in \mathscr{L}$, then the sequence $0 \to H \xrightarrow{i} G \xrightarrow{\pi} G/H \to 0$ (where i is the injection and π is the quotient map) is a PSE sequence in $\mathscr{L}$.

We now prove the basic fact about proper short exact sequences [from Moskowitz (1967)].

9.14 Proposition Let $0 \to G_1 \xrightarrow{f} G_2 \xrightarrow{g} G_3 \to 0$ be a PSE sequence in $\mathscr{L}$. Then the dual sequence $0 \to \hat{G}_3 \xrightarrow{g^*} \hat{G}_2 \xrightarrow{f^*} \hat{G}_1 \to 0$ is also a PSE sequence in $\mathscr{L}$.

Proof: By P.23(d) both g^* and f^* are proper. By P.23(b), g^* is one-one and f^* has dense image. Since $\text{im } f^*$ must be closed [P.15 or P.23(d)], we see that f^* is surjective. It remains only to show that $\text{im } g^* = \ker f^*$. Now by P.23(d) we have $\text{im } g^* = A(\hat{G}_2, \ker g)$. But exactness of the original sequence gives $\ker g = \text{im } f$, so $\text{im } g^* = A(\hat{G}_2, \text{im } f) = \ker f^*$ (cf. beginning of P.23). ∎

As is well known, the "injectives" of A are the divisible groups (Fuchs 1970, §21), while the "projectives" of A are the free abelian groups (Fuchs 1970, §14.6). Let us examine the corresponding situation in $\mathscr{L}$.

9.15 Definition A group $I \in \mathscr{L}$ is said to be *injective* iff, whenever $0 \to G_1 \xrightarrow{f} G_2 \xrightarrow{g} G_3 \to 0$ is a PSE sequence in $\mathscr{L}$ and $\phi \in \text{Hom}(G_1, I)$, there exists $\bar{\phi} \in \text{Hom}(G_2, I)$ such that $\bar{\phi} \circ f = \phi$, i.e., the following diagram commutes:

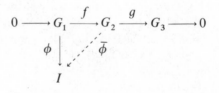

9.16 Remark It is easily verified that $I \in \mathscr{L}$ is injective iff, whenever $G \in \mathscr{L}$ and H is a closed subgroup of G, then any $\phi \in \text{Hom}(H, I)$ can be extended to $\bar{\phi} \in \text{Hom}(G,I)$. This formulation of injectivity, perhaps more concrete than Definition 9.15, is the one used by Dixmier (1957).

We come now to the determination of the injective LCA groups. The following result was first proved by Dixmier (1957) and later by Moskowitz (1967) in a different way. Our proof uses Theorem 9.12.

9.17 **Theorem** A group $I \in \mathscr{L}$ is injective iff I is a universal splitting group. Hence by 9.12, I is injective iff $I \cong R^n \times T^m$, for some $n \in Z^{+0}$ and cardinal number m.

Proof: Assume first that I is injective. Let $G \in \mathscr{L}$ contain a closed subgroup $H \cong I$ and let ϕ denote any topological isomorphism from H onto I. Consider the diagram

$$0 \longrightarrow H \xrightarrow{\ i\ } G \xrightarrow{\ \pi\ } G/H \longrightarrow 0$$
$$\phi \downarrow$$
$$I$$

where i is the injection and π the quotient map. Since the top row is a PSE sequence in $\mathscr{L}$, the injectivity of I implies the existence of $\bar{\phi} \in \text{Hom}(G,I)$ such that $\phi = \bar{\phi} \circ i$. Setting $f = \phi^{-1} \circ \bar{\phi}$ we see that $f \in \text{Hom}(G,H)$ and f is the identity on H. Therefore by 6.6, H splits from G. It follows that I is a universal splitting group. Conversely, suppose that I is a universal splitting group, so that I has the form $R^n \times T^m$ by Theorem 9.12. It is then evident from Remark 9.16 and the proof of Theorem 6.16 that I is injective. ∎

We now examine the dual concept of projectivity.

9.18 **Definition** A group $P \in \mathscr{L}$ is said to be *projective* iff, whenever $0 \to G_1 \xrightarrow{f} G_2 \xrightarrow{g} G_3 \to 0$ is a PSE sequence in $\mathscr{L}$ and $\psi \in \text{Hom}(P,G_3)$, there exists $\bar{\psi} \in \text{Hom}(P,G_2)$ such that $\psi = g \circ \bar{\psi}$, i.e., the following diagram commutes:

$$0 \longrightarrow G_1 \xrightarrow{\ f\ } G_2 \xrightarrow{\ g\ } G_3 \longrightarrow 0$$
$$\bar{\psi} \nwarrow \qquad \uparrow \psi$$
$$P$$

The next result [from Moskowitz (1967)] is naturally to be expected.

9.19 Proposition A group $P \in \mathscr{L}$ is projective iff $\hat{P}$ is injective.

Proof: The argument is a pleasant (and straightforward) exercise in duality, using 9.14 and appropriate parts of P.23. To the reader belong the details. ∎

The characterization of the projective groups [from Moskowitz (1967)] is now readily obtained.

9.20 Corollary A group $P \in \mathscr{L}$ is projective iff $P \cong R^n \times Z^{m^*}$ for some $n \in Z^{+0}$ and cardinal number $\mathfrak{m}$.

Proof: Just combine Theorem 9.17 and the preceding proposition. ∎

Miscellanea

9.21 (Grove and Lardy 1971) Call $G \in \mathscr{L}$ an SQ-group iff G has the following property: $f \in \text{Hom}(G, G)$ is one-one iff $f(G)$ is dense in G. The groups R, Q, $\hat{Q}$, and F_p are SQ-groups. More generally, a torsion-free divisible LCA group G is an SQ-group iff $G \cong R^n \times Q^m \times (\hat{Q})^k \times E$, where n, m, k are in Z^{+0} and E is the minimal divisible extension of a product of various groups J_p, where, for a given $p \in \mathscr{P}$, only finitely many J_p's occur. (See 9.6.)

9.22 Call an LCA group H a *universal quasi-splitting group* (UQS group for short) iff whenever $G \in \mathscr{L}$ has a closed subgroup $A \cong H$ then G also has a closed subgroup $A' \cong H$ such that A' splits from G. Obviously, a universal splitting group is a UQS group. The converse, however, fails. Indeed, $Z(p^\infty)$ turns out to be a UQS group (see 6.4 for an illustration of this phenomemon). The group Q is not a UQS group, since it is easily seen that if G is as in 6.17 we cannot have $G \cong Q \times G_0$ for any closed subgroup G_0 of G. Perhaps surprisingly, however, the group $Q^{\aleph_0^*}$ is a UQS group. This is a special case of the following interesting theorem of Ahern and Jewett (1965) (who use the term "universal external direct factor" for what we call a UQS group): A group $H \in \mathscr{L}$ is a UQS group iff $H \cong R^n \times T^m \times D_1 \times D_2$, where $n \in Z^{+0}$, $\mathfrak{m}$ is a cardinal number, $D_1 \in \mathscr{L}_d$ is a divisible torsion group, and $D_2 \cong Q^{\mathfrak{n}^*}$, where either $\mathfrak{n} = 0$ or $\mathfrak{n} \geq \aleph_0$.

9.23 Let A, B, and C be in $\mathscr{L}$ and let $f \in \text{Hom}(A, B)$ and $g \in \text{Hom}(B, C)$ be proper. Then $g \circ f$ need not be proper. If, however, f is surjective or g is injective, then $g \circ f$ is easily seen to be proper. For this and several other results (some rather deep) about proper homomorphisms, see Fulp and Griffith (1971).

9.24 (Moskowitz 1967) Let $I \in \mathscr{L}$ be injective for the class of compactly generated LCA groups (i.e., I satisfies Definition 9.15 with the stipulation that G_1, G_2, and G_3 be compactly generated). Then I is injective.

9.25 (Moskowitz 1967) Let G be an LCA group.

(a) G is compactly generated iff G has a short injective resolution in $\mathscr{L}$, i.e., iff there exist injective groups I_1 and I_2 in $\mathscr{L}$ and a PSE sequence $0 \to G \to I_1 \to I_2 \to 0$.

(b) Dually, G is an NSS group iff G has a short projective resolution in $\mathscr{L}$, i.e., iff there exist projective groups P_1 and P_2 in $\mathscr{L}$ and a PSE sequence $0 \to P_1 \to P_2 \to G \to 0$.

9.26 (Moskowitz 1967) As we know [P.4(d)], a subgroup of a free abelian group is again free. This generalizes to LCA groups as follows: A closed subgroup of a projective LCA group is again projective. (It is perhaps easiest to dualize and use 9.9.)

9.27 *Envoi* In this chapter we have only scratched the surface of what is known about the homological and categorical aspects of $\mathscr{L}$. It would take us too far afield to explore the subject in the depth which it deserves, but the following paragraphs should serve to indicate some of its various aspects not yet mentioned.

The first full-scale homological treatment of LCA groups is to be found in Moskowitz (1967). Here Moskowitz defines a discrete group-valued Ext functor for certain restricted classes of LCA groups and proves a number of results about this functor. Later, Fulp and Griffith (1971) generalized Moskowitz's construction in the following way. First, an elaborate machinery is developed designed primarily to show that certain homomorphisms, arising naturally in various commutative diagrams, are proper. (Our Proposition 6.5 is a by-product of this.) The stage is now set for the definition of their extension functor. A PSE sequence $E : 0 \to A \overset{\alpha}{\to} B \overset{\beta}{\to} C \to 0$ in $\mathscr{L}$ is said (naturally enough) to be an *extension of A by C*. If $E' : 0 \to A \overset{\alpha'}{\to} B' \overset{\beta'}{\to} C \to 0$ is another extension of A by C we say that E

is *congruent* to E' iff there exists $\phi \in \text{Hom}(B,B')$ such that the following diagram (wherein i denotes the identity) commutes:

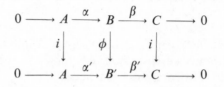

It then turns out (as a consequence of the elaborate machinery already mentioned) that ϕ must be a topological isomorphism from B onto B', so that congruence becomes a symmetric relation; reflexivity and transitivity of congruence are clear. Denote by $\text{Ext}(C,A)$ the set of congruence classes of extensions of A by C. Fulp and Griffith then prove a technical lemma which allows them to employ standard homological techniques to turn $\text{Ext}(C,A)$ into a discrete abelian group. (Ext turns out to have the same functorial properties as the usual Ext functor in abelian group theory; moreover, the two Ext's agree as groups when A and C are both discrete.) The zero of $\text{Ext}(C,A)$ is the congruence class of the "split" extension $0 \to A \overset{i}{\to} A \times C \overset{\pi}{\to} C \to 0$, where i is the injection and π is the projection. It is straightforward to verify that an extension $0 \to A \overset{\alpha}{\to} B \overset{\beta}{\to} C \to 0$ is congruent to this split extension iff $\alpha(A)$ splits from B.

Fulp and Griffith then proceed to establish a number of interesting and important properties of Ext, such as $\text{Ext}(C,A) \simeq \text{Ext}(\hat{A},\hat{C})$ for all A and C in $\mathscr{L}$, and the validity of the analogue of Theorem §51.3 of Fuchs (1970), by which the functors Hom and Ext are connected by a pair of long exact sequences. They are then able to conclude that $\text{Ext}(C,A) = \{0\}$ under various conditions imposed on A and C, which is of course just the thing that a "splitter" is looking for. For example, our Theorem 9.12 may be phrased: $\text{Ext}(C,A) = 0$ for all $C \in \mathscr{L}$ iff A has the form $R^n \times T^m$ for some $n \in Z^{+0}$ and cardinal number m. Among the results of Fulp and Griffith (1971) we mention the following (with a slight rephrasing): If $A \in \mathscr{L}$ then $\text{Ext}(C,A) = 0$ for all totally disconnected $C \in \mathscr{L}$ iff A is a d-group (that is, the d-groups are the only LCA groups which always split from containing LCA groups whenever the quotient is totally disconnected).

In Fulp (1972) the program initiated in Fulp and Griffith (1971) is carried further. Among several interesting results we cite the following. By analogy with the cotorsion groups of abelian group theory [see Fuchs

(1970, §54)] Fulp calls an LCA group A an $\mathscr{L}$-*cotorsion* group iff $\text{Ext}(C,A)$ $= \{0\}$ for every torsion-free $C \in \mathscr{L}$. He then shows that the connected LCA groups are $\mathscr{L}$-cotorsion and that a compact group $A \in \mathscr{L}$ is $\mathscr{L}$-cotorsion iff A is connected. Actually, parts of this have already appeared in Chapter 6 in nonhomological guise. For example, the argument of 6.20 can be used to show that a compact connected group $A \in \mathscr{L}$ is $\mathscr{L}$-cotorsion. Fulp also shows that a group $A \in \mathscr{L}_d$ is $\mathscr{L}$-cotorsion iff A is a divisible torsion group. (One direction of this may be quickly deduced from 6.32.) We note further that Proposition 6.23 implies that F_p is $\mathscr{L}$-cotorsion. Indeed, consider a PSE sequence $0 \to F_p \overset{\alpha}{\to} B \overset{\beta}{\to} C \to 0$ in $\mathscr{L}$, where C is torsion-free. It is easily seen that B must also be torsion-free, so $\alpha(F_p)$ splits from B by 6.23, i.e., $\text{Ext}(C,F_p) = 0$ and F_p is accordingly $\mathscr{L}$-cotorsion. Fulp obtains necessary and sufficient conditions for a group to be $\mathscr{L}$-cotorsion, but, as he points out, these conditions await further clarification.

Fulp (1970) also studied pure PSE sequences in $\mathscr{L}$ [i.e., PSE sequences of the form $0 \to A \overset{\alpha}{\to} B \overset{\beta}{\to} C \to 0$ in which $\alpha(A)$ is a pure subgroup of B] by way of analogy with the functor Pext of abelian group theory. Unfortunately, his Proposition 2, on which much of the paper is based, is not valid for all LCA groups (see Example 7.2 for a counterexample). As far as the author is aware, many of the interesting problems addressed in this paper remain unsolved. In connection with t-pure exact sequences (cf. 7.23) the reader should consult Khan (1973c).

The functors Hom and Tor and the tensor product defined on certain subclasses of $\mathscr{L}$ are studied in Moskowitz (1967). Further investigations of the tensor product may be found in Hofmann (1964) and Garling, (1966). Various technical complications arise in studying these functors in $\mathscr{L}$, and it seems safe to say that the subject has considerable room for further investigation.

The interested reader will also want to consult the work of Roeder on the category $\mathscr{L}$. Roeder (1971) is a study of the functorial aspects of Pontryagin duality. Roeder (1974) gives a very quick and elegant proof of the duality theorem by categorical arguments. A very readable treatment of an analogous approach to the duality theorem may be found in Morris (1977).

For a detailed cohomological study of compact abelian groups, the reader should consult Hofmann and Mostert (1973).

10

Different Topologies

> *"Like, but oh how different!"* [Wordsworth,
> *Poems of the Imagination, xxix*]

Many abelian groups can of course be endowed with several different locally compact group topologies, and the general question of how these different topologies are related has several fascinating ramifications. Though the subject is, comparatively speaking, quite an extensive one, we shall restrict our attention in the main body of this chapter (other topics being relegated to the Miscellanea) to just three questions:

(I) Which are the nondiscrete groups $G \in \mathscr{L}$ having the property that the only strictly stronger locally compact group topology on G is the discrete topology?

(II) Do the continuous characters of an LCA group determine its topology?

(III) If G is an LCA group and G_s denotes G with a strictly stronger locally compact group topology, how much larger is $(G_s)\hat{}$ than $\hat{G}$?

A complete answer to (I) was given (independently of one another) by Rajagopalan (1968b) and Rickert (1967). We shall present this as Theorem 10.1 and use it to tie up a loose end in Chapter 5. An affirmative answer to (II) first appears in Glicksberg (1962), followed shortly after by an independent and quite different solution in Varopoulos (1964a). Our plan of attack for answering (II) and (III) is as follows. We shall first give an answer to (III) (see Proposition 10.6) which is admittedly incomplete, but sufficient to prove the affirmative answer to (II), which we then give as Theorem 10.7. We then go on to give a far sharper answer to (III). Question III has an interesting history. Let G and G_s be as in our formulation of (III). Hewitt (1963) proved [independently of Glicksberg (1962)] not only that G_s must have at least one continuous character which is not continuous on G, but that in fact there must be $2^{\aleph_1}$ such characters, where $\aleph_1$ is the least uncountable cardinal. [Later, Ross (1965) gave a different and shorter proof of this.] If we assume the continuum hypothesis then of course we can say that G acquires 2^c extra continuous characters when its topology is strengthened. Hewitt (1963) conjectured that this striking conclusion could be reached without invoking the continuum hypothesis. His conjecture was subsequently proved independently (and by similar means) by three authors, namely, Rajagopalan (1964), Ross (1964), and Varopoulos (1964b). By blending ingredients of the proofs of Rajagopalan and Ross we shall prove this as Theorem 10.13. Readers who prefer to assume the continuum hypothesis may thus ignore these added complications, but the delicate analysis required to avoid the continuum hypothesis has interest in its own right.

Now we tackle question I. Let $G \in \mathscr{L}$ be nondiscrete and let G_s denote G with a strictly stronger locally compact group topology. Hewitt (1963) showed that if G is R or T, then G_s is discrete. Indeed, let i be the continuous (but not open) identity map from R_s onto R. Just as in the proof of 5.15 we conclude that either R_s is discrete or else $R_s \cong R$. However, the latter cannot occur, for if so P.30(b) would imply that i is open. Thus $R_s = R_d$. The argument for T is quite similar and is left to the reader. It is also true that if $G = J_p$ for any $p \in \mathscr{P}$ then G_s is discrete. For if $i : (J_p)_s \to J_p$ is the continuous (but not open) identity map, the argument of 5.20 shows that either $(J_p)_s$ is discrete or else $(J_p)_s \cong J_p$. But the latter cannot occur, since if so the compactness of J_p would force i to be open. Can we find groups G other than R, T, and J_p such that G_s must be discrete? The groups $R \times R_d$ and F_p have this property; however, we note that $R \times R_d$ (resp. F_p) contains R (resp. J_p) as an open subgroup. In fact, it is easily proved from P.30(a) that if G contains a copy of R, T, or J_p as an open subgroup, then G_s is necessarily discrete. Basing our argument on Rajagopalan (1968b) and Rickert (1967), we now prove the converse.

10.1 Theorem Let $G \in \mathscr{L}$ be nondiscrete. Then G has the property that the only strictly stronger locally compact group topology on G is the discrete topology iff G contains a copy of R, T, or J_p (for some prime p) as an open subgroup.

Proof: Assume that any strictly stronger locally compact group topology on G must be discrete, and let F be any closed subgroup of G. We claim that F is either open or discrete. For if not, define a new topology τ on

G by taking as a neighborhood base for 0 the neighborhoods of 0 in F (with its relative topology). It is easily seen that this new group (G, τ) is locally compact and that τ is at least as strong as the original G-topology. But if F is neither open nor discrete, then τ is strictly stronger than the G-topology but is not the discrete topology. This contradiction proves our claim that each closed subgroup of G is either open or discrete. In particular, $c(G)$ is either open or trivial.

Assume first that $c(G)$ is open but not trivial. Since no proper closed subgroup of $c(G)$ can be open, it follows from the first paragraph that each proper closed subgroup of $c(G)$ is discrete, whence by 1.4 either $c(G) \cong R$ or $c(G) \cong T$. Thus in this case G has R or T as an open subgroup.

Before proceeding to the case $c(G) = \{0\}$, let us note the following simple fact. If K is a group of the form $\Pi_{i \in I} K_i$, where each K_i is a nontrivial compact group, I is an infinite index set, and K has the (compact) product topology, then K contains closed subgroups which are neither open nor discrete.

Now, finally, we assume that G is totally disconnected, and we let K be any compact open subgroup of G. Since G is not discrete, K must be infinite. Now if K were a torsion group, P.24(b) would imply that K had the form described in the preceding paragraph, so that K (hence G) would contain a closed subgroup which is neither open nor discrete, contradicting the first paragraph. Therefore K must have some element x of infinite order. Let A be the compact monothetic group $\overline{\mathrm{gp}(x)}$. By 5.5(e) we have $A \cong \Pi_{p \in \mathscr{P}} A_p$, where for each $p \in \mathscr{P}$ we have $A_p \cong J_p$ or $A_p \cong Z(p^n)$ for some $n \in Z^{+0}$. If none of the A_p's were J_p then (since A is infinite) we would again, by the preceding paragraph, have a contradiction. Hence $A_p \cong J_p$ for some prime p. Since J_p is hardly discrete, we conclude from the first paragraph that G contains a copy of J_p as an open subgroup. This completes the proof in one direction. Since the converse has already been indicated, we are done. ∎

See 10.14 for a generalization of Theorem 10.1.

We are now in a position to prove the assertion mentioned in Chapter 5 (just before the Miscellanea) to the effect that the monothetic, solenoidal, and p-thetic groups are united by a common and exclusive property. The precise result is from Armacost and Armacost (1972a). (We here fill in a small gap in the proof.)

10.2 Corollary Let H be a noncompact group in $\mathscr{L}$. The following are equivalent:

(a) Any H-dense group $G \in \mathscr{L}$ is either compact or is topologically isomorphic to H.

(b) H is topologically isomorphic with Z, R, or $Z(p^\infty)$ for some $p \in \mathscr{P}$.

Proof: That (b) $\Rightarrow$ (a) follows from P.25, 5.15, and 5.20. Conversely, assume (a) and let L denote the nondiscrete group $\hat{H}$. From P.23(b) we see that L has the following property (a*): If $G \in \mathscr{L}$ and $f \in \mathrm{Hom}(G, L)$ is one-one, then either $G \cong L$ or else G is discrete. We must show that L is either T, R, or J_p for some $p \in \mathscr{P}$.

From P.29 and (a*) we see that either $L \cong R$ or else L has a compact open subgroup K. Assume the latter. Since L is not discrete, neither is K. Let $i : K \to L$ be the injection. Since K is not discrete, property (a*) shows that $L \cong K$. In particular, L must be compact. Since L is then either compact or topologically isomorphic to R, we can at least say this much: L is σ-compact.

Now let L_s denote L equipped with a strictly stronger locally compact group topology and let $i : L_s \to L$ be the continuous (but not open) identity mapping. By property (a*) we have either $L_s \cong L$ or L_s is discrete. But $L_s \cong L$ cannot occur, since if so L_s would, by the preceding paragraph, be σ-compact, whence by P.30(b), i would be open. (Note: Without σ-compactness it is conceivable that L_s and L could actually be topologically isomorphic, though not of course by i; see 10.16.) Hence L_s must be discrete. It now follows from Theorem 10.1 that L_s contains an open subgroup U topologically isomorphic to T, R, or some J_p. Since U is not discrete, property (a*) implies that $L \cong U$. This completes the proof that (a) $\Rightarrow$ (b). ∎

In order to expedite our investigation of questions II and III we shall need the following result, which is a direct consequence of the important theorem of Kakutani (1943) which we have stated as P.24(d).

10.3 Lemma Let $G \in \mathscr{L}$ be compact and infinite. Then G has 2^m discontinuous characters, where $m = 2^{|\hat{G}|}$.

Proof: The number of characters (discontinuous or not) on G is just $|(G_d)\hat{\ }|$. By P.24(d) applied twice we have $|(G_d)\hat{\ }| = 2^{|G_d|} = 2^{|G|} = 2^m$. Since only $|\hat{G}|$ characters on G are continuous and $2^m > |\hat{G}|$, there must be 2^m characters on G that are discontinuous. ∎

We note that P.24(d) and hence 10.3 do not use the continuum hypothesis.

We now begin to answer question III. The following special result is adapted from Hewitt (1963).

10.4 Lemma Let G be an abelian group which becomes a compact group under a topology τ. Let τ' be a strictly stronger locally compact group topology on G. Then there is a character γ of G which is τ'-continuous but not τ-continuous.

Proof: Let $i:(G,\tau') \to (G,\tau)$ be the continuous (but not open) identity map. By P.29 we know that (G,τ') either has a closed subgroup $A \cong R^n$ for some $n \in Z^+$ or else has a compact open subgroup K.

Assume first that (G,τ') has a closed subgroup $A \cong R^n$ where $n \in Z^+$. If $i(A)$ were closed in (G,τ) then P.30(b) would imply that $i(A) \cong R^n$, violating the compactness of (G,τ). Hence $i(A)$ is not closed, so if H denotes the τ-closure of $i(A)$ we have $H \supsetneq i(A)$. Thus if we set $B = i^{-1}(H)$ we have that B is τ'-closed and $B \supsetneq A$. We may then find [P.21(a)] a τ'-continuous character $\gamma_0 \neq 1$ of B which annihilates A. Now γ_0 is the restriction of some τ'-continuous character γ of G. But γ cannot be τ-continuous, since γ annihilates $i(A)$ but not its τ-closure H. This completes the proof in this case.

Now suppose that (G,τ') has a compact open subgroup K. Now $i(K)$ is not open in (G,τ), since if so P.30(a) would imply that i is open. But $i(K) = K$ is compact and hence closed in (G,τ), so G/K in the τ-quotient topology belongs to $\mathscr{L}$ and is in fact compact and nondiscrete. By 10.3, G/K has many characters which are discontinuous in this topology. All we need is one, say γ_0. Let π be the quotient map from G onto G/K. Define a character γ of G by $\gamma = \gamma_0 \circ \pi$. It is straightforward to verify that γ cannot be τ-continuous; γ is, however, τ'-continuous, since γ annihilates the τ'-open subgroup K of G. This completes the proof. ∎

The next proposition [from Hewitt (1963)] not only leads us to one of our goals but also has considerable interest in its own right. Our proof is different in some particulars from Hewitt's. For a far-reaching generalization the reader should consult Rajagopalan (1968b).

10.5 Proposition Let G denote R^n (for some $n \in Z^+$) equipped with a strictly stronger locally compact group topology. Then $G \cong R^a \times (R_d)^{n-a}$ for some integer a such that $0 \leq a < n$. It follows that G has 2^c continuous characters which are not continuous in the usual topology of R^n.

Proof: By P.29 we may write $G = A \oplus B$, where $A \cong R^a$ for some

$a \in Z^{+0}$ and B contains a compact open subgroup K. Let $i : G \to R^n$ be the continuous (but not open) identity mapping and let f be a topological isomorphism from R^a onto A. Define $g \in \text{Hom}(R^a, R^n)$ by $g = i \circ f$. One verifies easily (by using the density of the rationals in the reals) that g must be a linear mapping in the vector space sense, considering R^a and R^n as vector spaces over R. Since then $g(R^a)$ is a subspace of R^n and since g is one-one, we have $a \leq n$. Now $a = n$ cannot occur, since if so g would be surjective; this would force B to be $\{0\}$, in which case we would have $G = A \cong R^n$, whence by P.30(b), i would be open. Therefore $a < n$. It remains only to show that $B \cong (R_d)^{n-a}$. But $(R_d)^{n-a}$ is just the group $(Q^c)^*$, so by P.9(e) we need only verify that B is a discrete torsion-free divisible group of cardinality c. Now $b(R^n) = \{0\}$, whence $b(G) = \{0\}$. In particular, K is trivial, so B is discrete. It is obvious that B is torsion-free and divisible and that $|B| \leq c$. It remains then to show that $|B| \geq c$. However, the quotient space $R^n/g(R^a)$ is an $(n - a)$-dimensional vector space over R and thus has cardinality c. Moreover, it is directly verified that the mapping $\phi : B \to R^n/g(R^a)$, defined by $\phi(b) = i(b) + g(R^a)$ for each $b \in B$, is surjective. Thus $|B| \geq c$, and we are done, except, of course, to remark that the last statement of the proposition follows immediately from the fact [P.24(d)] that $|(R_d)\hat{\ }| = 2^c$. ∎

We now arrive at our first stopping point in our solution of question III. The result is due to Kaplansky (unpublished) and first appears in Glicksberg (1962). It is a special case of the main result of Hewitt (1963). Proofs may also be found in Varopoulos (1964a) and Ross (1965).

10.6 Proposition Let G be a group which becomes an LCA group under a topology τ. Let τ' be a strictly stronger locally compact group topology on G. Then there is a τ'-continuous character γ of G which is not τ-continuous.

Proof: By P.29, (G, τ) contains an open subgroup $U = A \oplus B$, where $A \cong R^n$ for some $n \in Z^{+0}$ and B is compact. Let $\tau(A)$ denote the topology that A inherits from (G, τ), with similar meanings for $\tau'(A)$, $\tau(B)$, and $\tau'(B)$. Clearly $\tau'(A) \supseteq \tau(A)$ and $\tau'(B) \supseteq \tau(B)$. We first show that either $\tau'(A) \neq \tau(A)$ or $\tau'(B) \neq \tau(B)$. Assume not, and let $\iota : (G, \tau) \to (G, \tau')$ be the identity map. Let $\{x_i\}_{i \in I}$ be any net in G which is τ-convergent to 0 (here I is some directed set). Then x_i is eventually in U, say for $i \geq i_0 \in I$. For such i we can write $x_i = a_i + b_i$, with $a_i \in A$ and $b_i \in B$. Then $a_i \overset{\tau}{\to} 0$ and $b_i \overset{\tau}{\to} 0$. But if $\tau'(A) = \tau(A)$ and $\tau'(B) = \tau(B)$ we have $a_i \overset{\tau'}{\to} 0$ and $b_i \overset{\tau'}{\to} 0$, whence $x_i \overset{\tau'}{\to} 0$. We conclude that ι is continuous at 0 and therefore on all of

(G,τ). But this implies that $\tau' \subseteq \tau$, a contradiction. Therefore either (1) $\tau'(A) \supsetneq \tau(A)$ or (2) $\tau'(B) \supsetneq \tau(B)$. If (1) holds we note that $\tau'(A)$ is a strictly stronger locally compact group topology on A $[\cong R^n$ in $(G,\tau)]$, whence by the last part of Proposition 10.5 we can find a $\tau'(A)$-continuous character γ_0 of A which is not $\tau(A)$-continuous. By P.21(a) we may then extend γ_0 to some $\gamma \in (G,\tau')\hat{\,}$, and it is evident that γ cannot be τ-continuous. If (2) holds we invoke Lemma 10.4 to reach the same conclusion. ∎

We now know that when an LCA group has its topology strengthened to make a new LCA group, new continuous characters must arise. By using this fact we can give an affirmative answer to question II. This affirmative answer appears first in Glicksberg (1962), which makes use of certain compactness theorems from functional analysis. (Glicksberg also mentions an alternative approach based on an unpublished Baire category argument due to Kaplansky.) Shortly afterwards Varopoulos (1964a) gave an independent and quite different argument, in which the problem is reduced to the situation covered by the preceding proposition (which, however, he handles differently). In our proof we make this reduction by adapting a key idea (that of the union of two locally compact group topologies) of Varopoulos (1964a). (Although we regard Theorem 10.7 below as a *terminus ad quem*, it does follow from a slightly more general result; see 10.17.)

10.7 *Theorem* Let G be an abelian group and let τ_1 and τ_2 be two locally compact group topologies on G. If $(G,\tau_1)\hat{\,} = (G,\tau_2)\hat{\,}$ then $\tau_1 = \tau_2$.

Proof: Set $G_1 = (G,\tau_1)$ and $G_2 = (G,\tau_2)$ and write G' for $G_1 \times G_2$ with the product topology. Let D be the diagonal of $G':D = \{(x,y) \in G': x = y\}$. We show first that D is closed in G'. Indeed, let $\{(x_i,x_i)\}_{i \in I}$ be a net in D converging to some $(y,z) \in G'$ (here I is some directed set). We must show that $y = z$. If $y \neq z$ then we can find some $\gamma \in \hat{G}_1$ such that $\gamma(y) \neq \gamma(z)$ [P.19(a)]. Now $x_i \to y$ in the topology τ_1, so $\gamma(x_i) \to \gamma(y) \in T$. But $x_i \to z$ in the topology τ_2, and since by hypothesis γ is also τ_2-continuous we have $\gamma(x_i) \to \gamma(z)$ in T. But then $\gamma(y) = \gamma(z)$, a contradiction. Thus D is closed in G', and since D is clearly a subgroup of G' we see that D becomes an LCA group in its topology inherited from G'.

We are now going to define a new topology on G. Let f be the map from G onto D defined by $f(x) = (x,x)$ for each $x \in G$. The collection of sets $\{f^{-1}(U):U$ open in $D\}$ forms a topology on G which we denote by τ, and it is evident $(G,\tau) \cong D$, so that (G,τ) becomes an LCA group. Moreover, we have $\tau \supseteq \tau_1$ and $\tau \supseteq \tau_2$. For let V_1 be open in (G, τ_1). We have $V_1 = f^{-1}((V_1 \times G_2) \cap D)$. But since $(V_1 \times G_2) \cap D$ is open in D we see by the definition of τ that V_1 is τ-open. Hence $\tau \supseteq \tau_1$, and similarly $\tau \supseteq \tau_2$.

Now let γ be any τ-continuous character of G. Define a character ψ on D by $\psi(x,x) = \gamma(x)$ for each $(x,x) \in D$. It is evident that ψ is continuous on D. We may now extend ψ to a continuous character $\bar{\psi}$ on G'. By P.19(b) we see that there exist $\gamma_1 \in \hat{G}_1$ and $\gamma_2 \in \hat{G}_2$ such that $\bar{\psi}(x,y) = \gamma_1(x)\gamma_2(y)$ for each $(x,y) \in G'$. Hence $\gamma = \gamma_1 \cdot \gamma_2$ on G. Now since γ_2 is τ_2-continuous and $\hat{G}_2 = \hat{G}_1$ by hypothesis, we know that γ_2 is also τ_1-continuous. Since γ_1 is τ_1-continuous and $\gamma = \gamma_1 \cdot \gamma_2$, we conclude that γ is τ_1-continuous. In summary: $\tau \supseteq \tau_1$, but every τ-continuous character γ of G is also τ_1-continuous. It follows from Proposition 10.6 that $\tau = \tau_1$. The same argument gives $\tau = \tau_2$, whence we have $\tau_1 = \tau_2$, as desired. ∎

At this point we begin a series of results designed to obviate the use of the continuum hypothesis in our final answer to question III. (Readers who hold the continuum hypothesis to be true may wish to pass directly to Proposition 10.12.) The first of these is adapted from Rajagopalan (1964).

10.8 Lemma Let $G \in \mathscr{L}$ be compact and let $n \in Z^+$. Let $f \in \mathrm{Hom}(R^n, G)$ be one-one and have dense image and let $f^* \in \mathrm{Hom}(\hat{G}, \widehat{(R^n)})$ be the adjoint of f. Let ψ denote the topological isomorphism from R^n onto $\widehat{(R^n)}$ defined in this way: If $\bar{r} = (r_1, \ldots, r_n) \in R^n$ then $\psi(\bar{r})$ is the character of R^n which evaluated at any $\bar{s} = (s_1, \ldots, s_n) \in R^n$ is $\exp i(r_1 s_1 + \cdots + r_n s_n)$ [cf. P.19(b) and P.20(b)]. Set $\bar{f} = \psi^{-1} \circ f^* \in \mathrm{Hom}(\hat{G}, R^n)$. Then there exist $\gamma_1, \ldots, \gamma_n$ in $\hat{G}$ such that

(a) $\bar{f}(\gamma_1), \ldots, \bar{f}(\gamma_n)$ are linearly independent in R^n (when R^n is regarded as a vector space over R).

(b) $\hat{G}/H$ is infinite, where H is the subgroup of $\hat{G}$ generated by $\gamma_1, \ldots, \gamma_n$.

Proof: By P.23(b) we see that f^* has dense image, so $\bar{f}(\hat{G})$ is dense in R^n. Let $\mathrm{sp}\,\bar{f}(\hat{G})$ denote the subspace of R^n spanned by $\bar{f}(\hat{G})$. Since subspaces of R^n are closed we have $R^n = \overline{\bar{f}(\hat{G})} \subseteq \overline{\mathrm{sp}\,\bar{f}(\hat{G})} = \mathrm{sp}\,\bar{f}(\hat{G}) \subseteq R^n$, whence $\mathrm{sp}\,\bar{f}(\hat{G}) = R^n$. Therefore we can find n linearly independent vectors in $\bar{f}(\hat{G})$, say $\bar{f}(\gamma_1), \ldots, \bar{f}(\gamma_n)$. Let $H = \mathrm{gp}\{\gamma_1, \ldots, \gamma_n\} \subseteq \hat{G}$. If $\hat{G}/H$ is infinite, we are of course finished. If $\hat{G}/H$ is finite, then $\hat{G}$ is finitely generated. Now $\hat{G}$ is torsion-free [since f^* is one-one by P.23(b)], so by P.6(b) we have $\hat{G} \cong Z^r$ for some positive integer r. But $\bar{f}(\hat{G})$ cannot be dense in R^n if $r \leq n$, so we have $r > n$. Let $\gamma_1', \ldots, \gamma_r'$ be independent generators of $\hat{G}$. Since sp $\overline{f}(\hat{G}) = R^n$ we see that $\overline{f}(\gamma_1'), \ldots, \overline{f}(\gamma_r')$ span R^n, whence we may select a basis, which we denote by $\overline{f}(\gamma_1), \ldots, \overline{f}(\gamma_n)$. (Here each γ_i is one of the γ_j'.) Letting $H = \mathrm{gp}\{\gamma_1, \ldots, \gamma_n\}$ we conclude from the fact that $n < r$ that $\hat{G}/H$ is infinite, whence the result. ∎

We shall also need the following special result.

10.9 Lemma Let the vectors $\alpha_1, \ldots, \alpha_n$ form a basis for the vector space R^n. Then the set A of all $\gamma \in (R^n)\hat{\ }$ which annihilate $\alpha_1, \ldots, \alpha_n$ is countable.

Proof: Let $\varepsilon_1, \ldots, \varepsilon_n$ be the standard unit vectors of R^n. Define a (continuous) linear transformation $\phi : R^n \to R^n$ by requiring $\phi(\alpha_i) = \varepsilon_i$ for $i = 1, \ldots, n$. Also define $\psi \in \mathrm{Hom}(R^n, T^n)$ by $\psi(\beta) = (\exp(2\pi i r_1), \ldots, \exp(2\pi i r_n))$, where $\beta = r_1\varepsilon_1 + \cdots + r_n\varepsilon_n$ for unique $r_1, \ldots, r_n$ in R. Setting $f = \psi \circ \phi$ we see that f is a continuous homomorphism from R^n onto T^n. Since f is open [P.30(b)] we have $R^n/\ker f \cong T^n$. Thus $Z^n \cong (T^n)\hat{\ } \cong A((R^n)\hat{\ }, \ker f)$. But it is easily checked that $\ker f$ is the closed subgroup H of R^n consisting of all integral linear combinations of $\alpha_1, \ldots, \alpha_n$. Now evidently $A = A((R^n)\hat{\ }, H)$, so $A \cong Z^n$ and in particular $|A| = \aleph_0$. ∎

The next result [from Rajagopalan (1964)] is of independent interest. A closely similar result is to be found in Varopoulos (1964b).

10.10 Proposition Let $G \in \mathscr{L}$ be compact and let $n \in Z^+$. Suppose that $f \in \mathrm{Hom}(R^n, G)$ is one-one and has dense image. Then G has a subgroup B such that $|B| \geq \mathfrak{c}$ and $G = f(R^n) + B$. In particular, we have $|G/\mathrm{im}\, f| \geq \mathfrak{c}$.

Proof: Since $f(R^n)$ is divisible we can find [by P.9(d)] a subgroup B of G such that $G = f(R^n) + B$. All we need do is show that $|B| \geq \mathfrak{c}$. To this end find $\gamma_1, \ldots, \gamma_n$ in $\hat{G}$ satisfying conditions (a) and (b) of Lemma 10.8. Set $S = A(G, H)$, so that $x \in G$ belongs to S iff $\gamma_i(x) = 1$ for $i = 1, \ldots, n$. We have $S \cong (\hat{G}/H)\hat{\ }$, whence by P.24(d), $|S| = 2^{|\hat{G}/H|} \geq \mathfrak{c}$. Now suppose that $f(\overline{r}) \in S$ for some $\overline{r} \in R^n$. Then $\gamma_i(f(\overline{r})) = 1$ for $i = 1, \ldots, n$. A brief computation shows that $\gamma_i(f(\overline{r}))$ is just $\psi(\overline{r})$ evaluated at $\overline{f}(\gamma_i)$. Thus $\psi(\overline{r})$ annihilates the n linearly independent vectors $\overline{f}(\gamma_1), \ldots, \overline{f}(\gamma_n)$, whence by Lemma 10.9 we have $|f(R^n) \cap S| \leq \aleph_0$. But since $|S| \geq \mathfrak{c}$ we must have $|S/(f(R^n) \cap S)| \geq \mathfrak{c}$. Let $E \subseteq S$ be a set consisting of exactly one representative of each coset of $S/(f(R^n) \cap S)$. Then of course $|E| \geq \mathfrak{c}$. We show that $|B| \geq \mathfrak{c}$ by exhibiting a one-one map ϕ from E into B. Let π be the projection from $G = f(R^n) + B$ onto B and let ϕ be the restriction of π to E. Suppose that $\phi(e_1) = \phi(e_2)$ for two elements e_1 and e_2 of E. Write $e_i = f(\bar{r}_i) + b_i$, where $\bar{r}_i \in R^n$ and $b_i \in B$ for $i = 1, 2$. We have $\phi(e_1) = \pi(e_1) = b_1$ and $\phi(e_2) = \pi(e_2) = b_2$, whence $b_1 = b_2$, so that $e_1 - e_2 = f(\bar{r}_1 - \bar{r}_2) \in f(R^n)$. Since e_1 and e_2 are both in the subgroup S of G we have $e_1 - e_2 \in f(R^n) \cap S$. But then e_1 and e_2 both belong to the same coset of $S/(f(R^n) \cap S)$, so by the definition of E we have $e_1 = e_2$. Thus ϕ is one-one, and we are done. ∎

The following consequence is from Ross (1964).

10.11 Corollary Let F and G be compact groups in $\mathscr{L}$ and let $n \in Z^+$. Suppose $f \in \mathrm{Hom}(R^n \times F, G)$ is one-one and has dense image in G. Then $|G/\mathrm{im}\, f| \geq \mathfrak{c}$.

Proof: Let $f(F)$ denote the compact subgroup $f(\{0\} \times F)$ of G. Define a mapping $\bar{f}: R^n \to G/f(F)$ by the rule $\bar{f}(\bar{r}) = f(\bar{r}, 0) + f(F)$ for each $\bar{r} \in R^n$. It is clear that $\bar{f}$ is a one-one continuous homomorphism whose image is dense in the compact group $G/f(F)$. Therefore by Proposition 10.10 we have $|(G/f(F))/\mathrm{im}\, \bar{f}| \geq \mathfrak{c}$. But it is clear that $\mathrm{im}\, \bar{f}$ is the subgroup $\mathrm{im}\, f/f(F)$ of $G/f(F)$. Therefore the second isomorphism theorem for groups (cf. §2.2 of [HR]) says that $|G/\mathrm{im}\, f| \geq \mathfrak{c}$. ∎ [Note: Ross also requires that f have proper image in G; this, however, in light of P.30(b), is a consequence of the other hypotheses.]

Having now done the technical work to get around the continuum hypothesis, we come to our penultimate result [from Ross (1964)], which leads right away to our final answer to question III.

10.12 Proposition Let G be an abelian group with two locally compact group topologies τ and τ'. If τ' is strictly stronger than τ then $|(G, \tau')\widehat{}| \geq 2^{\mathfrak{c}}$.

Proof: Step I. Assume that $(G, \tau) \cong R^n$ for some $n \in Z^+$. By P.24(d) we have $|(R_d)\widehat{}| = 2^{\mathfrak{c}}$, whence by 10.5 we have $|(G, \tau')\widehat{}| = 2^{\mathfrak{c}}$ as well.

Step II. Assume that (G, τ) is compact, and let $i : (G, \tau') \to (G, \tau)$ be the continuous (but not open) identity mapping. By P.29, (G, τ') contains an open subgroup $H = V \oplus K$, where $V \cong R^n$ for some $n \in Z^{+0}$ and K is compact. Suppose we knew that $|G/H| \geq \mathfrak{c}$. It would then follow from P.24(d) that the dual of the discrete group G/H had cardinality $\geq 2^{\mathfrak{c}}$, whence we could conclude that $|(G, \tau')\widehat{}| \geq 2^{\mathfrak{c}}$. Let us then complete this step by showing in fact that $|G/H| \geq \mathfrak{c}$.

If $|G/H| \leq \aleph_0$ then (G, τ') would be σ-compact, whence [P.30(b)] i

would be open, a contradiction. Therefore $|G/H| > \aleph_0$, and those readers assuming the continuum hypothesis already know that $|G/H| \geq \mathfrak{c}$. Otherwise, we proceed as follows. If $V = \{0\}$ then $H = K$ is compact in (G, τ'), whence H is compact and hence closed in (G, τ). Thus the infinite group G/H is compact in its τ-quotient topology, whence by P.24(d) we have $|G/H| \geq \mathfrak{c}$. If on the other hand $V \neq \{0\}$, then of course $n \geq 1$. Let g be a topological isomorphism from $R^n \times K$ onto H and let $\bar{H}$ denote the τ-closure of H in G. Define a map $f : R^n \times K \to \bar{H}$ by $f = i \circ g$. Since $\bar{H}$ is compact and f satisfies the conditions of 10.11, we find that $|\bar{H}/\mathrm{im}\, f| \geq \mathfrak{c}$. But this just says that $|\bar{H}/H| \geq \mathfrak{c}$, whence $|G/H| \geq \mathfrak{c}$ too. Therefore in this case we have $|(G, \tau')\hat{\,}| \geq 2^{\mathfrak{c}}$.

Step III. We know by P.29 that (G, τ) contains an open subgroup of the form $A \oplus B$, where $A \cong R^n$ for some $n \in Z^{+0}$ and B is compact. Just as in the proof of Proposition 10.6 we find (using the notation adopted there) that either $\tau'(A) \supsetneq \tau(A)$ or $\tau'(B) \supsetneq \tau(B)$. In the former case we see from Step I that $|(A, \tau'(A))\hat{\,}| = 2^{\mathfrak{c}}$ whence by P.21(a) we have $|(G, \tau')\hat{\,}| \geq 2^{\mathfrak{c}}$. In the latter case we apply Step II to reach the same conclusion. ∎

We can at last prove our final answer to question III. As mentioned earlier, this was conjectured by Hewitt (1963) and proved independently by Rajagopalan (1964), Ross (1964), and Varopoulos (1964b). (Varopoulos gives a slightly stronger answer; see 10.19.)

10.13 Theorem Let G be an LCA group under two topologies τ and τ'. If τ' is strictly stronger than τ there exist at least $2^{\mathfrak{c}}$ characters of G which are τ'-continuous but not τ-continuous.

Proof: Let N denote the set of all characters γ of G which are τ'-continuous but not τ-continuous. We must show that $|N| \geq 2^{\mathfrak{c}}$. By Proposition 10.6 we know that N is not empty. We distinguish two cases. If $|(G, \tau)\hat{\,}| \geq 2^{\mathfrak{c}}$ and $\gamma_0 \in N$, it is clear that $\gamma\gamma_0 \in N$ for each $\gamma \in (G, \tau)\hat{\,}$, whence $|N| \geq |(G, \tau)\hat{\,}| \geq 2^{\mathfrak{c}}$. Assume on the other hand that $|(G, \tau)\hat{\,}| < 2^{\mathfrak{c}}$. By Proposition 10.12 we have $|(G, \tau')\hat{\,}| \geq 2^{\mathfrak{c}}$. But since $(G, \tau')\hat{\,}$ is the disjoint union of $(G, \tau)\hat{\,}$ and N, we have $|(G, \tau')\hat{\,}| = |(G, \tau)\hat{\,}| + |N|$, whence $|N| \geq 2^{\mathfrak{c}}$. ∎

Miscellanea

10.14 (Rickert 1967) Let G be any LCA group under a topology τ and let $\mathscr{T}(G)$ denote the set of all locally compact group topologies τ' on G such that $\tau' \supseteq \tau$. Then if $|\mathcal{T}(G)| < \mathfrak{c}$ we have $|\mathcal{T}(G)| < \aleph_0$. Moreover, one can list eight types of LCA groups (involving R, T, various groups J_p, and various solenoidal groups) having the property that a group $G \in \mathcal{L}$ satisfies $|\mathcal{T}(G)| < \mathfrak{c}$ iff G contains an open subgroup topologically isomorphic to one of the types listed. For each type, $|\mathcal{T}(G)|$ can be computed (and turns out to be a power of 2 in each case). See Rickert's paper for the details.

10.15 Let G be an LCA group under two topologies τ_1 and τ_2. Let $[\tau_1, \tau_2]$ denote the set of all locally compact group topologies τ on G such that $\tau_1 \subseteq \tau \subseteq \tau_2$. By means of a lengthy and ingenious argument Janakiraman and Rajagopalan (1973) prove that either $\|[\tau_1, \tau_2]\| < \aleph_0$ or $\|[\tau_1, \tau_2]\| \geq \mathfrak{c}$, thus confirming a conjecture of Rickert (1967). A deep and detailed study of the case $\|[\tau_1, \tau_2]\| < \aleph_0$ may be found in Miller and Rajagopalan (1975). The main result is that in this case $\|[\tau_1, \tau_2]\|$ must be a power of 2; in fact, $[\tau_1, \tau_2]$ has the same lattice structure as that of the power set of some finite set ordered by inclusion.

10.16 Let $G \in \mathcal{L}$ and let G_s denote G endowed with a strictly stronger locally compact group topology. It may happen that $G_s \cong G$. [Set $G = \Pi_{n=1}^{\infty} K_n$, where $K_1 = (\hat{Q})_d$, $K_n = \hat{Q}$ for $n = 2, 3, \ldots$, and set $G_s = \Pi_{n=1}^{\infty} K_n'$, where $K_1' = K_2' = (\hat{Q})_d$ and $K_n' = \hat{Q}$ for $n = 3, 4, \ldots$. Observe that the identity map $i : G_s \to G$ is continuous but not open. But $(\hat{Q})_d \cong Q^{c^*} \cong Q^{c^*} \times Q^{c^*} \cong (\hat{Q})_d \times (\hat{Q})_d$, whence we can conclude $G_s \cong G$.]

10.17 (Varopoulos 1964a) Let G be an abelian group and let τ_1 and τ_2 be two locally compact group topologies on G. If every τ_1-continuous character on G is also τ_2-continuous, then $\tau_1 \subseteq \tau_2$. (The proof of Theorem 10.7 is easily adapted to cover this situation.)

10.18 Let τ and τ' be locally compact group topologies on an abelian group G. The following are equivalent: (a) $\tau' \supsetneq \tau$ and (b) $(G, \tau)\hat{}$ is a proper dense subgroup of $(G, \tau')\hat{}$. [The implication (a) $\Rightarrow$ (b) is from Hewitt (1963): Just look at i^*, where $i : (G, \tau') \to (G, \tau)$ is the identity map, and use 10.6. The converse follows from 10.17.]

10.19 (Varopoulos 1964b) Let G, τ, and τ' be as in 10.13. Then $|(G, \tau')\hat{}/(G, \tau)\hat{}| \geq 2^{\mathfrak{c}}$.

10.20 (Varopoulos 1965) Let G, τ, and τ' be as in 10.13. There exists a sequence $\{x_n\}_{n=1}^{\infty}$ of elements of G and an element x_0 of G such that $x_n \to x_0$

in τ but not in τ'. [The proof of this is rather complicated. Varopoulos uses the result to show that the size of $(G,\tau)\hat{}$ is, in a measure-theoretic sense, small in comparison to that of $(G,\tau')\hat{}$. A consequence is that if $G \in \mathscr{L}$ is noncompact, then G is of outer (Haar) measure zero in its Bohr compactification $\beta(G)$.]

10.21 Let G be an abelian group and let τ_1 and τ_2 be locally compact group topologies for G. If (G,τ_1) and (G,τ_2) have the same closed subgroups, what can be said about the relation between τ_1 and τ_2? Ross (1965) gives a simple example to show that τ_1 need not equal τ_2. Nevertheless, in this example we still have $(G,\tau_1) \cong (G,\tau_2)$. Two questions arise: (1) Under what circumstances can we conclude that $\tau_1 = \tau_2$, and (2) Do we always have $(G,\tau_1) \cong (G,\tau_2)$? Ross (1965) shows that (G,τ_1) is compact iff (G,τ_2) is compact. He then goes on to show that if (G,τ_1) is compact and totally disconnected, then $\tau_1 = \tau_2$. This result has been improved independently by Rajagopalan (1968b) and Rickert (1967): If either (G,τ_1) or (G,τ_2) is totally disconnected, then $\tau_1 = \tau_2$. Both Rajagopalan (1968b) and Rickert (1967) shed considerable light on questions (1) and (2). The interested reader should also consult Baker (1968). As far as the author is aware, definitive answers to questions (1) and (2) are not yet available.

10.22 (Scheinberg 1974) Let G_1 and G_2 be connected LCA groups. If G_1 and G_2 are homeomorphic as topological spaces, then $G_1 \cong G_2$. (Scheinberg uses methods not developed in this book. He also proves other results on the relations between homeomorphisms and isomorphisms.)

10.23 The algebraic structure of all those abelian groups which admit compact group topologies is completely known (see §25.25 of [HR]). Some abelian groups, of course, admit several compact topologies; see, for example, Hawley (1971) for a detailed discussion of the compact group topologies on the real numbers. Other groups, however, are not so obliging. Hulanicki (1964) contains several interesting and delicate results on the possible compact topologies on abelian groups, and in his Theorem 8.10 Hulanicki characterizes those abelian groups which admit exactly one compact group topology [see also Orsatti (1970)]. Among these groups are the p-adic integer groups J_p. Soundararajan (1969) proved further that the groups J_p admit no other nondiscrete locally compact group topologies. [This was later re-proved by Corwin (1976) using a different argument based on his structure theorem mentioned after 9.6.]

[Soundararajan's argument is essentially as follows. Let $p \in \mathscr{P}$ be fixed. Now the endomorphisms of J_p are just the multiplications of J_p [see Fuchs (1970, p. 181)]. Hence if $\phi \neq 0$ is any endomorphism of J_p, then im ϕ is compact and therefore open in J_p [P.18(d)], so im ϕ has finite index in J_p. Now let G be any LCA group algebraically isomorphic to J_p. Since G is reduced, we conclude from P.29 and P.28(b) that G has a compact, totally disconnected open subgroup K. If $K = \{0\}$ then G is discrete. Otherwise we use P.24(a) to show that K contains a closed subgroup $H \cong J_p$. Let ϕ be any algebraic isomorphism from G onto H. Then ϕ is a nontrivial endomorphism of G, so $H = $ im ϕ has finite index in G. Therefore G is compact. The proof can now be completed by using P.11.] Soundararajan (1969) then goes on to characterize all the compact groups (G, τ) in $\mathscr{L}$ such that the only locally compact group topologies on G are τ and the discrete topology. The structure of general groups (G, τ) in $\mathscr{L}$ having this same property seems to be unknown [see Soundararajan (1971)].

10.24 Very few abelian groups G have the property that they become compact or locally compact topological groups under the "natural" topology formed by using the sets $\{nG\}_{n=1}^{\infty}$ as a base for the neighborhoods of 0. See Orsatti (1967) for a characterization of these groups.

10.25 Peterson (1973) studies the relation between large subgroups (that is, subgroups of finite index) of a group G and the continuity of characters on G (cf. 5.41). Among several interesting results we cite the following: Let G be any LCA group. Let $\mathfrak{m}$ be the number of large nonopen subgroups of G and let $\mathfrak{n}$ be the number of discontinuous characters of G having finite order. Then if either $\mathfrak{m}$ or $\mathfrak{n}$ is infinite we have $\mathfrak{m} = \mathfrak{n}$. The paper also lists some open questions, one of which concerns the result cited.

References

Ahern, P. R., and R. I. Jewett (1965). Factorization of locally compact abelian groups. Ill. J. Math. *9*, 230–235. (MR *31*, 3536)

Anzai, H., and S. Kakutani (1943). Bohr compactifications of a locally compact abelian group I & II. Proc. Imp. Acad. Tokyo *19*, 476–480 and 533–539. (MR *7*, 374)

Armacost, D. L. (1970). Well-known LCA groups characterized by their closed subgroups. Proc. Amer. Math. Soc. *25*, 625–629. (MR *41*, 5544)

———(1971a). Can an LCA group be anti-self-dual? Proc. Amer. Math. Soc. *27*, 186–188. (MR *42*, 1938)

———(1971b). Sufficiency classes of LCA groups. Trans. Amer. Math. Soc. *158*, 331–338. (MR *45*, 819)

———(1972). Mapping properties of characters of LCA groups. Fund. Math. *76*, 1–7. (MR *46*, 7441)

———(1971c). Generators of monothetic groups. Can. J. Math. *23*, 791–796. (MR *44*, 2885)

———(1973). Remark on my paper "Generators of monothetic groups." Can. J. Math. *25*, 672. (MR *47*, 5170)

———(1974). On pure subgroups of LCA groups. Proc. Amer. Math. Soc. *45*, 414–418. (MR *49*, 10813)

———(1976). Compactly cogenerated LCA groups. Pac. J. Math. *65*, 1–12. (MR *58*, 16944)

———On a theorem of Baayen and Helmberg (to appear).

Armacost, D. L., and W. L. Armacost (1972a). On *p*-thetic groups. Pac. J. Math. *41*, 295–301. (MR *48*, 8680)

———(1972b). On *Q*-dense and densely divisible LCA groups. Proc. Amer. Math. Soc. *36*, 301–305. (MR *46*, 5527)

———(1978). Uniqueness in structure theorems for LCA groups. Can. J. Math. *30*, 593–599. (MR *58*, 28254)

Armacost, D. L., and R. R. Bruner (1973). Locally compact groups without distinct isomorphic closed subgroups. Proc. Amer. Math. Soc. *40*, 260–264. (MR *48*, 8681)

Baayen, P. C., and G. Helmberg (1965). On families of equi-uniformly distributed sequences in compact spaces. Math. Ann. *161*, 255–278. (MR *32*, 5627)

Baker, J. W. (1968). A note on the duality of locally compact groups. Glasgow Math. J. *9*, 87–91. (MR *38*, 3375)

Bourbaki, N. (1967). Éléments de Mathématique. Fascicule XXXII: Théories Spectrales. Hermann, Paris.

Braconnier, J. (1948). Sur les groupes topologiques localement compacts. J. Math. Pures Appl., N. S. *27*, 1–85. (MR *10*, 11)

Bruner, R. R. (1972). Radicals and torsion theories in locally compact ˙ abelian groups. Unpublished.

See also Armacost, D. L.

Čarin, V. S. (1967). Groups with complemented subgroups (Russian). Dokl. Akad. Nauk SSSR *173*, 50–53. [English version in Sov. Math. *8*, 344–347 (1967).] (MR *35*, 287)

Comfort, W. W., and G. L. Itzkowitz (1977). Density character in topological groups. Math. Ann. *226*, 223–227. (MR *56*, 5767)

Comfort, W. W., and K. A. Ross (1964). Topologies induced by groups of characters. Fund. Math. *55*, 283–291. (MR *30*, 183)

Corwin, L. (1970). Some remarks on self-dual locally compact abelian groups. Trans. Amer. Math. Soc. *148*, 613–622. (MR *42*, 4670)

———(1976). Uniqueness of topology for the *p*-adic integers. Proc. Amer. Math. Soc. *55*, 432–434. (MR *54*, 2865)

Dietrich, W., Jr. (1972). Dense decompositions of locally compact groups. Colloq. Math. *24*, 147–151. (MR *47*, 6935)

Dixmier, J. (1957). Quelques propriétés des groupes abéliens localement compacts. Bull. Sci. Math. (2) *81*, 38–48. (MR *20*, 3926)

Eckmann, B. (1944). Über monothetische Gruppen. Comment. Math. Helv. *16*, 249–263. (MR *6*, 146)

Enochs, E. (1964). Homotopy groups of compact abelian groups. Proc. Amer. Math. Soc. *15*, 878–881. (MR *29*, 6493)

Fan, K. (1970). On local connectedness of locally compact abelian groups. Math. Ann. *187*, 114–116. (MR *41*, 7022)

Flor, P. (1978). Eine Bemerkung über lokalcompakte abelsche Gruppen. Fund. Math. *101*, 135–136. (MR *80*b, 22008)

Fuchs, L. (1960). Abelian Groups. Pergamon Press, London.
––– (1970). Infinite Abelian Groups (Vol. I). Academic Press, New York.
––– (1973). Infinite Abelian Groups (Vol. II). Academic Press, New York.
Fulp, R. O (1970). Homological study of purity in locally compact groups. Proc. London Math. Soc. *21*, 501–512. (MR *43*, 4952)
––– (1972). Splitting locally compact abelian groups. Mich. Math. J. *19*, 47–55. (MR *45*, 3629)
Fulp, R. O., and P. A. Griffith (1971). Extensions of locally compact abelian groups I and II. Trans. Amer. Math. Soc. *154*, 341–363. (MR *42*, 7751)
Garling, D. (1966). Tensor products of topological abelian groups. J. Reine Angew. Math. *223*, 164–182. (MR *33*, 7453)
Glicksberg, I. (1962). Uniform boundedness for groups. Can. J. Math. *14*, 269–277. (MR *27*, 5856)
Griffith, P. A. *See* Fulp, R. O.
Grove, L. C., and L. J. Lardy (1971). A finiteness condition for locally compact abelian groups. J. Austral. Math. Soc. *12*, 115–121. (MR *43*, 814)
Guichardet, A. (1968). Analyse Harmonique Commutative. Dunod, Paris.
Halmos, P. R. (1944). Comment on the real line. Bull. Amer. Math. Soc. *50*, 877–878. (MR *6*, 145)
Halmos, P. R., and H. Samelson (1942). On monothetic groups. Proc. Nat. Acad. Sci. U.S.A. *28*, 254–258. (MR *4*, 2)
Hartman, S., and A. Hulanicki (1955). Les sous-groupes purs et leurs duals. Fund. Math. *45*, 71–77. (MR *19*, 1063)
––– (1958). Sur les ensembles denses de puissance minimum dans les groupes topologiques. Colloq. Math. *6*, 187–191. (MR *21*, 94)
Hartman, S., and C. Ryll-Nardzewski (1957). Zur Theorie der lokal-kompakten abelschen Gruppen. Colloq. Math. *4*, 157–188. (MR *19*, 430)
Hawley, D. (1971). Compact topologies for *R*. Proc. Amer. Math. Soc. *30*, 566–572. (MR *43*, 7548)
Helmberg, G. *See* Baayen, P. C.
Hewitt, E. (1963). A remark on characters of locally compact Abelian groups. Fund. Math. *53*, 55–64. (MR *27*, 4889)
Hewitt, E., and K. Ross (1963). [HR] Abstract Harmonic Analysis (Vol. I). Springer-Verlag, Berlin, and Academic Press, New York.
Hofmann, K. H. (1964). Tensorprodukte lokal kompakter abelscher

Gruppen. J. Reine Angew. Math. *216*, 134–149. (MR *29*, 4834)

Hofmann, K. H., and P. S. Mostert (1973). Cohomology Theories for Compact Abelian Groups. Springer-Verlag, New York.

Hulanicki, A. (1964). Compact Abelian groups and extensions of Haar measures. Rozprawy Mat. *38*. (MR *31*, 270)

See also Hartman, S.

Husain, T. (1966). Introduction to Topological Groups. Saunders, Philadelphia.

Isiwata, T. (1955). On the connectedness of locally compact groups. Sci. Rep. Tokyo Kyoiku Daigaku, Sect. A5, 8–13. (MR *17*, 172)

Itzkowitz, G. (1968). The existence of homomorphisms in compact connected abelian groups. Proc. Amer. Math. Soc. *19*, 214–216. (MR *36*, 2734)

———(1972). On the density character of compact topological groups. Fund. Math. *75*, 201–203 (MR *46*, 3684)

See also Comfort, W. W.

Janakiraman, S., and M. Rajagopalan (1973). Topologies in locally compact groups II. Ill. J. Math. *17*, 177–197. (MR *47*, 3593)

Jewett, R. I. *See* Ahern, P. R.

Kakutani, S. (1943). On cardinal numbers related with a compact abelian group. Proc. Imp. Acad. Tokyo *19*, 366–372. (MR *7*, 375)

See also Anzai, H.

Kaplansky, I. (1969). Infinite Abelian Groups (revised edition). University of Michigan Press, Ann Arbor.

Keesling, J. (1972). The group of homeomorphisms of a solenoid. Trans. Amer. Math. Soc. *172*, 119–131. (MR *47*, 4284)

Khan, M. A. (1973a). The existence of certain homomorphisms in locally compact abelian groups. J. Nat. Sci. Math. (Government College, Lahore) *13*, 125–144. (MR *51*, 13124)

———(1973b). A notion of purity in locally compact groups. J. Sci. Univ. Karachi *2*, 128–130. (MR *55*, 5787)

———(1973c). A note on *t*-pure extensions of locally compact groups. J. Sci. Univ. Karachi *2*, 297–299.

———(1976). Generalized locally compact cyclic groups. J. Nat. Sci. Math. (Government College, Lahore) *16*, 61–68. (MR *57*, 12766)

———(1980). Chain conditions on subgroups of LCA groups. Pac. J. Math. *86*, 517–534.

Kuipers, L., and H. Niederreiter (1974). Uniform Distribution of Sequences. Wiley (Interscience), New York.

Lardy, L. J. *See* Grove, L. C.

Leptin, H. (1956). Eine Kennzeichnung der reinen Untergruppen abelscher Gruppen. Acta Math. Acad. Sci. Hungar. 7, 169–171. (MR 19, 290)

Levin, M. D. (1971). The automorphism group of a locally compact abelian group. Acta Math. 127, 259–278. (MR 54, 10481)

Mackey, G. W. (1946). A remark on locally compact Abelian groups. Bull. Amer. Math. Soc. 52, 940–944. (MR 8, 311)

Miller, C. B., and M. Rajagopalan (1975). Topologies in locally compact groups, III. Proc. London Math. Soc. 31, 55–78. (MR 51, 10525)

Morris, S. A. (1972). Locally compact abelian groups and the variety of topological groups generated by the reals. Proc. Amer. Math. Soc. 34, 290–292. (MR 45, 3630)

———(1977). Pontryagin Duality and the Structure of Locally Compact Abelian Groups. Cambridge University Press, Cambridge.

Moskowitz, M. (1967). Homological algebra in locally compact abelian groups. Trans. Amer. Math. Soc. 127, 361–404. (MR 35, 5861)

Mostert, P. S. See Hofmann, K. H.

Muhin, Ju. N. (1967). Locally compact groups with a distributive lattice of closed subgroups (Russian). Sibirsk. Mat. Ž. 8, 366–375. (MR 35, 1697)

———(1970). Topological abelian groups with Dedekind lattice of closed subgroups (Russian). Mat. Zametki 8, 509–519. [English version in Math. Notes 8, 755–760 (1970).] (MR 44, 4143)

———(1976). The number of generating elements of a locally compact abelian group (Russian). Algebraic studies, pp. 23–33, 119. Izdanie Ural. Politehn. Inst., Sverdlovsk. (MR 56, 530)

Niederreiter, H. See Kuipers, L.

Olubummo, A., and M. Rajagopalan (1976). Anti-self-dual groups. Comment. Math. Prace Mat. 19, 111–112. (MR 54, 10474)

Orsatti, A. (1967). Una caratterizzazione dei gruppi abeliani compatti o localmente compatti nella topologia naturale. Rend. Sem. Mat. Univ. Padova 39, 219–225. (MR 37, 2893)

———(1970). Sui gruppi abeliani ridotti che ammettono una unica topologia compatta. Rend. Sem. Mat. Univ. Padova 43, 341–347. (MR 43, 7546)

Peterson, H. L. (1973). Discontinuous characters and subgroups of finite index. Pac. J. Math. 44, 683–691. (MR 47, 5174)

Pontryagin, L. S. (1934). The theory of topological commutative groups. Ann. of Math. (2) 35, 361–388.

———(1966). Topological Groups (second edition, trans. from the Russian by A. Brown). Gordon and Breach, New York.

Preston, G. C. (1956). On locally compact totally disconnected Abelian groups and their character groups. Pac. J. Math. *6*, 121–134. (MR *18*, 51)

Rajagopalan, M. (1964). Characters of locally compact abelian groups. Math. Z. *86*, 268–272. (MR *31*, 4852)

———(1968a). Structure of monogenic groups. Ill. J. Math. *12*, 205–214. (MR *37*, 5322)

———(1968b). Topologies in locally compact groups. Math. Ann. *176*, 169–180. (MR *37*, 335)

See also Janakiraman, S., Miller, C. B., and Olubummo, A.

Rajagopalan, M., and T. Soundararajan (1967). On self-dual LCA groups. Bull. Amer. Math. Soc. *73*, 985–986. (MR *36*, 614)

———(1969). Structure of self-dual torsion-free metric LCA groups. Fund. Math. *65*, 309–316. (MR *40*, 643)

Rajagopalan, M., and H. Subrahmanian (1976). Dense subgroups of locally compact groups. Colloq. Math. *35*, 289–292. (MR *54*, 5381)

Reiter, H. (1968). Classical Harmonic Analysis and Locally Compact Groups. Oxford University Press, Oxford.

Rickert, N. (1967). Locally compact topologies for groups. Trans. Amer. Math. Soc. *126*, 225–235. (MR *34*, 2770)

Robertson, L. C. (1967). Connectivity, divisibility and torsion. Trans. Amer. Math. Soc. *128*, 482–505. (MR *36*, 302)

———(c. 1968). Transfinite torsion, *p*-constituents, and splitting in locally compact abelian groups. Unpublished.

Robertson, L. C., and B. M. Schreiber (1968). The additive structure of integer groups and *p*-adic number fields. Proc. Amer. Math. Soc. *19*, 1453–1456. (MR *37*, 6934)

Roeder, D. W. (1971). Functorial characterizations of Pontryagin duality. Trans. Amer. Math. Soc. *154*, 151–176. (MR *43*, 4956)

———(1974). Category theory applied to Pontryagin duality. Pac. J. Math. *52*, 519–527. (MR *50*, 13365)

Ross, K. (1964). Locally compact groups and the continuum hypothesis. Colloq. Math. *13*, 21–25. (MR *30*, 2103)

———(1965). Closed subgroups of locally compact Abelian groups. Fund. Math. *56*, 241–244. (MR *30*, 2104)

See also Comfort, W. W. and Hewitt, E.

Rudin, W. (1962). Fourier Analysis on Groups. Wiley (Interscience), New York.

Ryll-Nardzewski, C. *See* Hartman, S.

Samelson, H. *See* Halmos, P. R.

Scheinberg, S. (1974). Homeomorphism and isomorphism of abelian groups. Can. J. Math. *26*, 1515–1519. (MR *50*, 4809)

Schreiber, B. M. *See* Robertson, L. C.

Shelah, S. (1974). Infinite Abelian groups, Whitehead problem and some constructions. Isr. J. Math. *18*, 243–256. (MR *50*, 9582)

Soundararajan, T. (1969). The topological group of the *p*-adic integers. Publ. Math. Debrecen *16*, 75–78.

———(1971). Some results on locally compact groups. General Topology and Its Relations to Modern Analysis and Algebra (Proceedings of the Kanpur Topological Conference, 1968), Publishing House of the Czechoslovak Academy of Sciences, Prague, and Academic Press, New York, pp. 297–298.

See also Rajagopalan, M.

Subrahmanian, H. *See* Rajagopalan, M.

Takamatsu, R. (1976). On polythetic groups. Proc. Japan Acad. *52*, 17–20. (MR *53*, 697)

Varopoulos, N. Th. (1964a). Studies in harmonic analysis. Proc. Cambridge Philos. Soc. *60*, 465–516. (MR *29*, 1284)

———(1964b). A theorem on cardinal numbers associated with a locally compact Abelian group. Proc. Cambridge Philos. Soc. *60*, 701–704. (MR *29*, 3570)

———(1965). A theorem on the Bohr compactification of a locally compact Abelian group. Proc. Cambridge Philos. Soc. *61*, 65–68. (MR *30*, 3173)

Venkataraman, R. (1970). On locally compact groups which are topologically pure in their Bohr compactifications. Fund. Math. *69*, 103–107. (MR *42*, 6519)

Vilenkin, N. (1946). Direct decompositions of topological groups I. Mat. Sb., N. S. *19 (61)*, 85–154. [English translation from the Russian by E. Hewitt in A.M.S. Translations, Series 1, Volume 8, Providence, Rhode Island (1962).]

Weil, A. (1941, 1951). L'Intégration dans les Groupes Topologiques et ses Applications. Hermann & Cie., Paris.

Whitehead, J.H.C. (1956). Duality in topology. J. London Math. Soc. *31*, 134–148. (MR *17*, 1118)

Notation

$\mathscr{A}$	class of abelian groups, 1
$\bar{A}$	closure of A, 5
A^c	complement of A, 5
$A(\hat{G}, S)$	annihilator of S in $\hat{G}$, 9
$A(G, S')$	annihilator of S' in G, 10
$b(G)$	compact elements of G, 7
$\beta(G)$	Bohr compactification of G, 13
$\mathfrak{c}$	power of continuum, 1
$c(G)$	connected component of 0 in G, 6
$d(G)$	maximal divisible subgroup of G, 4
$e(G)$	set of generators of G, 57
f^*	adjoint of f, 10
F_p	p-adic numbers, 7
G^*	minimal divisible extension of G, 4
$G!$	set of x in G such that $n!x \to 0$, 32
$\hat{G}$	character group of G, 8
G_d	G with the discrete topology, 5
$G^{\mathfrak{m}}$	direct product of $\mathfrak{m}$ copies of G, 2
$G^{\mathfrak{m}*}$	weak direct product of $\mathfrak{m}$ copies of G, 2
$G_{(n)}$	set of x in G such that $nx = 0$, 1
G_p	p-component of G, 24
gp (S)	subgroup generated by S, 2
gp (x)	subgroup generated by x, 2
Hom (G, H)	set of continuous homomorphisms from G to H, 6
$H + K$	subgroup generated by H and K, 2
$H \dotplus K$	algebraic direct sum of H and K, 3
$H \oplus K$	topological direct sum of H and K, 6

$H \times K$	direct product of H and K, 2
im f	image of f, 1
J_p	p-adic integers, 7
ker f	kernel of f, 1
$\mathscr{L}$	class of LCA groups, 5
LCA	locally compact abelian and Hausdorff, 5
$\mathscr{L}_d$	class of discrete abelian groups, 5
LP	local direct product, 6
nG	set of elements nx in G, 1
$n(G)$	nongenerators of G, 57
$o(x)$	order of x, 1
$\mathscr{P}$	set of prime numbers, 1
Π	direct product, 2
Π^*	weak direct product, 2
Q	rational numbers, 2
$\hat{Q}$	character group of Q, 9
R	real numbers, 7
$\mathscr{R}$	radical, 53
$r(G)$	rank of G, 3
$r_0(G)$	torsion-free rank of G, 3
$r_p(G)$	p-rank of G, 3
$\mathscr{S}(H)$	sufficiency class of H, 42
$\mathscr{S}^*(H)$	dual sufficiency class of H, 42
$\Sigma_{i \in I} H_i$	subgroup generated by subgroups H_i, 2
Σ_p	p-adic solenoid, 64
T	circle group, 7
$t(G)$	torsion subgroup of G, 2
$w(G)$	weight of G, 8
$\|X\|$	cardinal number of set X, 1
Z	the integers, 1
$Z+$	the positive integers, 1
$Z+0$	the nonnegative integers, 1
$Z(n)$	cyclic group of order n, 2
$Z(p^\infty)$	quasicyclic group, 2

Index